Discrete Probability

HARBRACE COLLEGE MATHEMATICS SERIES

Salomon Bochner and W. G. Lister, EDITORS

PUBLISHED TITLES

Calculus, Karel de Leeuw
Linear Algebra, Ross A. Beaumont
Discrete Probability, R. A. Gangolli and Donald Ylvisaker
Functions of Several Variables, John W. Woll, Jr.
Algebraic Topology: An Introduction, William S. Massey

Discrete Probability

R. A. GANGOLLI
UNIVERSITY OF WASHINGTON

DONALD YLVISAKER
UNIVERSITY OF WASHINGTON

Harcourt, Brace & World, Inc.

New York / Chicago / San Francisco / Atlanta

ISBN 0-15-517690-0

LIBRARY OF CONGRESS CATALOG CARD NUMBER: 67-14631

PRINTED IN THE UNITED STATES OF AMERICA

Foreword

The Harbrace College Mathematics Series has been undertaken in response to the growing demands for flexibility in mathematics curricula. This series of concise, single-topic textbooks is designed to serve two primary purposes: First, to provide basic undergraduate text materials in compact, coordinated units. Second, to make available a variety of supplementary textbooks covering single topics.

To carry out these aims, the series editors and the publisher have selected as the foundation of the series a sequence of six textbooks covering functions, calculus, linear algebra, multivariate calculus, theory of functions, and theory of functions of several variables. Complementing this sequence are a number of other volumes on such topics as probability, statistics, differential equations, topology, differential geometry, and complex functions.

By permitting more flexibility in the construction of courses and course sequences, this series should encourage diversity and individuality in curricular patterns. Furthermore, if an instructor wishes to devise his own topical sequence for a course, the Harbrace College Mathematics Series provides him with a set of books built around a flexible pattern from which he may choose the elements of his new arrangement. Or, if an instructor wishes to supplement a full-sized textbook, this series provides him with a group of compact treatments of individual topics.

An additional and novel feature of the Harbrace College Mathematics Series is its continuing adaptability. As new topics gain emphasis in the curricula or as promising new treatments appear, books will be added to the series or existing volumes will be revised. In this way, we will meet the changing demands of the instruction of mathematics with both speed and flexibility.

SALOMON BOCHNER

W. G. LISTER

Preface

This book is written for a first course in the elements of discrete probability. A view common to many textbooks written for such an introductory course is that the first course is but a prelude to a traditional study of statistics. As a result, these textbooks cover certain topics in continuous probability in addition to the elements of discrete, or finite, probability theory. This approach we find unsatisfactory, on two counts. First, the discussion of the elements of discrete probability is necessarily brief. But students taking a course in probability theory are usually prepared for an accurate course in discrete probability, and treating this subject cursorily means that an opportunity to teach it effectively is being thrown away. Second, a minimal understanding of continuous probability is predicated on a sound understanding of the calculus. And, although many students who take a course in probability theory have had a year's exposure to the calculus, most of them are not prepared to use it effectively as a tool.

For these reasons, we have chosen to limit this book to the elements of discrete probability. We have tried to be precise and comprehensive in our coverage of finite sample spaces and random variables defined on such sample spaces. This stance naturally leads in some instances to a more formal presentation than is customary in introductory textbooks; we have found, however, that students gain far more from precision than from vague intuitive discussion.

The treatment of random variables culminates in Chapter 8 in a proof of the weak law of large numbers for sequences of independent and identically distributed random variables. Up to that point, the subject matter is largely independent of the calculus. Indeed, in the unstarred sections of the book (which constitute the bulk of the material) only once is an analytical notion used and even then only in the shape of the simple fact, used in proving the weak law of large numbers, that $1/n \to 0$ as $n \to \infty$.

In the starred sections of the book, the normal and Poisson approximations to the binomial distribution are discussed briefly and a statement of the central limit theorem is given. While a thorough discussion of these topics does presuppose an understanding of the calculus, we have found that including them in an informal way makes the subject more interesting and need not introduce undue pedagogical difficulty.

We have provided a large number of problems so that the instructor will be able to select those that suit him. Some of the problems are more difficult than others or require an acquaintance with the calculus; these exercises are marked with a star.

This textbook, then, is written for a first course in probability theory. In a program for mathematics majors it is appropriate for a course at the sophomore or junior level, and it provides the necessary foundation for a course in probability and statistics. For students of business, biological science, social science, or engineering, it can be used to establish the background for a course in statistical methodology.

The book can be handled at a rather leisurely pace in a course of one-quarter duration if the starred sections are omitted; or it can be adapted to a less leisurely—but more interesting—one-quarter course if the first eight chapters are treated. The entire book, of course, can be studied in one semester, with some time left for other topics of the instructor's choice.

R. A. GANGOLLI
DONALD YLVISAKER

Seattle, Washington

Contents

CHAPTER THREE

CHAPTER FOUR

CHAPTER FIVE

Discrete Probability

CHAPTER ONE

Sets

1.1 Introduction

This chapter—a mathematical prelude to the study of probability—is a modest and beginning study of sets. Our choice of material from set theory has been guided very much by our own selfish aims, and, indeed, we will carry the entire body of it over into our study of probability in a thinly veiled disguise.

The first notions covered here are those of set and subset. Next, within the context of a so-called universal set, relations between sets and operations upon sets are considered. The final section of the chapter introduces the concept of a product set. The framework within which these matters pertain to the study of probability theory is a major concern of Chapter 2.

1.2 Sets

A *set* is any well-defined collection of objects, the objects being called the *elements* or *members* of the set. Here, a well-defined collection is taken to mean that it is possible to decide whether any object is or is not an element of the set. The collection of Shakespeare plays, of former presidents of the United States, of New England states, and of real numbers between 1 and 2 are examples of sets. *Macbeth*, George Washington, Vermont, and $\frac{3}{2}$ are, respectively, examples of members of these sets. In each instance, a precise rule is provided for deciding whether or not a given object is a member of the set.

Two sets are said to be *equal* if they consist of precisely the same elements. No two of the sets described above are equal. However, the set of New England states and the set of the states of Maine, Vermont, New Hampshire, Massachusetts, Rhode Island, and Connecticut are equal. The order in which these states have been named is unimportant, for another ordering (interchanging Maine and Vermont, for instance) will only produce a set that is equal to the other two.

It is often necessary to consider one or more sets without particular regard to their specific nature. When this is so, they will be designated by the letters A, B, C, etc. In the same vein, their elements will be written x, y, z, etc. To indicate the membership of an object in a set, we write $x \in A$ (read "x is in A") as an abbreviation for the statement "the object x is an element of the set A." Nonmembership is similarly denoted by $x \notin A$ (read "x is not in A"). The equality of two sets A and B is expressed by writing $A = B$, and the definition of equality may be rephased as follows: $A = B$ if and only if $x \in A$ implies $x \in B$ and conversely.

To specify a set A, the entire membership of A must be made explicit. There are two general procedures for doing this. The first is simply that of listing the members of A with the (implicit) agreement that it is a complete list. This method is adopted, for example, by the management of an office building when it wishes to identify the collection of tenants. Listings are given in the form $A = \{x_1, x_2, \ldots, x_n\}$, when A is the set of elements $x_1, x_2, \ldots, x_n$. When a set has very many members, this method may not be sensible, or even possible.

An alternative method for specifying the members of a set is to give a verbal description of the elements of the set. The agreement then is that any object satisfying this description is a member of the set and no other object is. This technique was used in an essential way when we talked of the set of real numbers between 1 and 2. Also, when we talked of the sets of Shakespeare plays, of former presidents of the United States, and of New England states we used this technique; in these latter instances it would have been possible to list all the members of the various sets, but it would have been a tedious task. Whenever it is appropriate, a set is given in the form $A = \{x | x$ satisfies the given description$\}$, which may be read "A is the set of objects x that satisfy the description." Thus, $A = \{x | x$ is a real number and $1 < x < 2\}$ is our formal way of specifying the set of real numbers between 1 and 2. The set of New England states could be written as {Maine, Vermont, New Hampshire, Massachusetts, Rhode Island, Connecticut} or as $\{x | x$ is a New England state$\}$.

Remarks: It is a necessary part of human communication that sets of objects be described. Thus, one might speak of the Chinese people, the Library of Congress, the courses given at a state university, the First Family, foreign cars, the programming of a television network, the vertebrates, women drivers, and so on. Hand in hand with this goes the possibility of misunderstanding—the set intended may not be the set conveyed. The examples listed above range from sets whose membership is well known—in which case misunderstanding is unlikely—to sets whose membership requires considerable clarification. When describing sets in this book, we will exercise as much care as can be comfortably accommodated.

EXERCISES 1.2

1. Give an alternative specification of the following sets by listing members.
 (a) $A = \{x|\ x$ is a member of the First Family$\}$.
 (b) $A = \{x|\ x$ is a chapter heading of this book$\}$.
 (c) $A = \{x|\ x$ is a primary color$\}$.
 (d) $A = \{x|\ x$ is the capital city of a New England state$\}$.
 (e) $A = \{x|\ x$ is a major television network$\}$.
 (f) $A = \{x|\ x$ is a real number for which $x^2 - 5x + 4 = 0\}$.

2. Give a descriptive specification of the following sets.
 (a) $A = \{c, d, e, f, g, a, b\}$.
 (b) $A = \{1, 2, 3, 4, 5, 6, 7, 8, 9, 10\}$.
 (c) $A = \{$Brown, Columbia, Cornell, Dartmouth, Harvard, Pennsylvania, Princeton, Yale$\}$.
 (d) $A = \{$Chrysler Corporation, General Motors Corporation, Ford Motor Company$\}$.
 (e) $A = \{$u, e, a, i, o$\}$.
 (f) $A = \{$Germany, Italy, Japan$\}$.

3. If a set A has exactly n members for some integer n, it is called a *finite* set. If A has more than n members for each integer n, it is called an *infinite* set. Which of the following are finite sets?
 (a) $A = \{x|\ x$ is a Beethoven symphony$\}$.
 (b) $A = \{x|\ x$ is a real number for which $2x = 7\}$.
 (c) $A = \{x|\ x$ is an odd integer$\}$.
 (d) $A = \{x|\ x$ is a word in this book$\}$.
 (e) $A = \{x|\ x$ is a pair of humans living today$\}$.

4. Group the following collection of sets together into blocks so that within a block of sets any set is equal to any other set, while no two sets belonging to different blocks are equal.
 $A_1 = \{x|\ x$ is a real number and $|x - \frac{1}{2}| = \frac{1}{2}\}$.
 $A_2 = \{x|\ x$ is a positive integer smaller than $2\}$.
 $A_3 = \{1, 2\}$.
 $A_4 = \{x|\ x$ is a real number and $x^2 - 3x + \frac{9}{4} = 0\}$.
 $A_5 = \{1\}$.
 $A_6 = \{2, 1\}$.
 $A_7 = \{x|\ x$ is a real number equal to its square$\}$.

1.3 Subsets

Suppose that A and B are two sets. The set A is said to be a *subset* of B (or A is *contained* in or *included* in B) if every element of A is also an element of B. This important relationship between two sets is written succinctly as $A \subset B$. Thus, $A \subset B$ means that $x \in A$ implies $x \in B$. On the other hand, if A is not contained in B, i.e., not every element of A is also an element of B, we write $A \not\subset B$. For two given sets A and B, there are four possible inclusion relationships: (1) $A \subset B$ and $B \not\subset A$, (2) $B \subset A$ and $A \not\subset B$, (3) $A \not\subset B$ and $B \not\subset A$, or (4) $A \subset B$ and $B \subset A$. If $A \subset B$ but $B \not\subset A$, we say that A is *properly* contained in B (there is one exception to this which is noted below). On the other hand, if $A \subset B$ and $B \subset A$, we say that the inclusion is *improper*. In this last case, $A = B$.

To limit the discussion suitably, we choose a fixed set Ω (a deviation in notation for emphasis) and subsequently consider only those objects x which are elements of Ω and only those sets A which are subsets of Ω. The set Ω so chosen is termed a *universal set.* No special restriction need be placed on the choice made, but it will suffice for our purposes to take Ω to be a finite set, say $\Omega = \{\omega_1, \omega_2, \ldots, \omega_n\}$. The fixed set Ω might, for example, be the set of former presidents of the United States, and the discussion would be limited to former presidents and subsets of the set of former presidents.

To specify a subset A of Ω we need, of course, only make clear which of the members of Ω are members of A and which are not. We can do this, at least formally, in the following way: we proceed down the list of elements $\omega_1, \omega_2, \ldots, \omega_n$, associating with each element one of the two symbols $\in$ and $\notin$. In this manner, we specify a subset A of Ω if it is agreed that A consists of precisely those elements associated with the symbol $\in$. Going through all such associations will produce all possible subsets of Ω; it is evident that Ω itself arises from the constant choice of $\in$ at each ω_i. Allowing only a single choice of $\in$ will lead to the subsets $\{\omega_1\}, \{\omega_2\}, \ldots, \{\omega_n\}$, which are called the *simple subsets* of Ω. There is also a subset A of Ω determined by $\omega_i \notin A$ for each $i = 1, 2, \ldots, n$. The subset A is then a subset having no members at all. This subset, which will prove to be very useful, is written as $\varnothing$ and is called the *empty set.*

Inasmuch as the condition in the definition of inclusion is vacuous when applied to sets $\varnothing$ and A, we find that $\varnothing \subset A$ for any set A. This inclusion is also called *improper.* Combining this inclusion with the improper inclusion $A \subset B$ and $B \subset A$, we find that the inclusion $A \subset B$ is improper if $A = \varnothing$ or $A = B$. Equivalently, it may be seen that

	$\varnothing$	A_1	A_2	Ω
ω_1	$\notin$	$\in$	$\notin$	$\in$
ω_2	$\notin$	$\notin$	$\in$	$\in$

FIGURE 1.1

$A \subset B$ is a proper inclusion if A consists of some, but not all, of the members of B.

Let us temporarily take the number of elements of Ω to be small enough so that all the subsets of Ω can be easily written out. If $\Omega = \{\omega_1, \omega_2\}$, the subsets are those given in Figure 1.1. This figure indicates that the subsets are $\varnothing$, $A_1 = \{\omega_1\}$, $A_2 = \{\omega_2\}$, and $\Omega = \{\omega_1, \omega_2\}$. In this case, A_1 and A_2 are the proper subsets of Ω and are also the simple subsets of Ω.

Similarly, if $\Omega = \{\omega_1, \omega_2, \omega_3\}$, the subsets are as shown in Figure 1.2. There are eight subsets of Ω of which A_1, A_2, A_3, A_4, A_5, and A_6 are proper. A_1, A_2, and A_3 are the simple subsets of Ω.

It might be guessed at this point that if $\Omega = \{\omega_1, \omega_2, \ldots, \omega_n\}$, there are 2^n subsets of Ω. This is in fact so (see Exercise 7 at the end of this section), and on this basis we observe that there are $2^n - 2$ proper subsets of Ω, n of which are simple subsets.

Now we suppose that A and B are two subsets of Ω and consider certain relationships that may or may not exist between A and B. First let us imagine classifying the elements of Ω with respect to their simultaneous membership or nonmembership in the subsets A and B. For an element x of Ω, there are four possibilities: (1) $x \in A$ and $x \in B$, (2) $x \in A$ and $x \notin B$, (3) $x \notin A$ and $x \in B$, or (4) $x \notin A$ and $x \notin B$. If there are *no* elements x of Ω for which $x \in A$ and $x \notin B$, then every x of Ω which is in A must also be in B. We know this to be the inclusion relation $A \subset B$. If, on top of this, there are no elements x of Ω for which $x \notin A$ and $x \in B$, then also $B \subset A$ and in fact $A = B$.

Now let us suppose rather that there are no elements x of Ω for which $x \in A$ and $x \in B$, i.e., A and B have no elements in common. In such a case, we say that A and B are *disjoint*. The empty set $\varnothing$ having no members at all is then disjoint from every set A.

	$\varnothing$	A_1	A_2	A_3	A_4	A_5	A_6	Ω
ω_1	$\notin$	$\in$	$\notin$	$\notin$	$\in$	$\in$	$\notin$	$\in$
ω_2	$\notin$	$\notin$	$\in$	$\notin$	$\in$	$\notin$	$\in$	$\in$
ω_3	$\notin$	$\notin$	$\notin$	$\in$	$\notin$	$\in$	$\in$	$\in$

FIGURE 1.2

If a subset A of Ω is given, the elements of Ω that do not belong to A comprise a subset of Ω that is called the *complement* of A in Ω and is denoted by A'. When it is clear what the universal set Ω is, we say A' is the complement of A (instead of "the complement of A in Ω"). Thus, $A' = \{x|\, x \notin A\}$. Note that in writing $\{x|\, x \notin A\}$ we again consider only those x that are in Ω so that this complementary relationship is relative to the universal set Ω. It is easy to see that $\varnothing$ is the complement of Ω ($\Omega' = \varnothing$) and that Ω is the complement of $\varnothing$ ($\varnothing' = \Omega$). More generally, the complement of the subset A' of Ω is A, i.e., $(A')' = A$.

Example 1.1

Suppose $\Omega = \{x|\, x$ is a book of the Bible$\}$. $A = \{x|\, x$ is an Old Testament book of the Bible$\}$ and $B = \{x|\, x$ is a New Testament book of the Bible$\}$ are clearly complementary subsets of Ω. $C = \{x|\, x$ is a book of the Bible written by John$\}$ is a subset of B, for every book written by John is in the New Testament. This of course means that no such book is in the Old Testament, so that A and C are disjoint. The same may be said of $D = \{x|\, x$ is one of the four Gospels of the Bible$\}$. C and D have one element in common, the Gospel according to John, but neither is contained in the other. $\{$The Book of Genesis$\}$ is a simple subset of Ω, is contained properly in A, and is disjoint from B, C, and D. $E = \{x|\, x$ is a book of the Bible written by Pontius Pilate$\}$ has no members so we may write $E = \varnothing$.

EXERCISES 1.3

1. Prepare a table of all subsets of $\Omega = \{\omega_1, \omega_2, \omega_3, \omega_4\}$. How many subsets of Ω have exactly two members? Exactly three members? Which subsets of Ω are complements of each other? Give three pairs of proper subsets of Ω that are disjoint and not complements of each other.

2. Referring to Figure 1.2, indicate the inclusion relations (if any) that exist between
   (a) A_2 and A_5. **(b)** A_3 and A_5. **(c)** A_4 and A_5.
   Which of these pairs are disjoint? Which are complementary?

3. Let $\Omega = \{x|\, x$ is a former president of the United States$\}$. Give descriptive specifications of four proper subsets of Ω.

4. Let $\Omega = \{x|\, x$ is a city in the United States with population in excess of 200,000$\}$. Give descriptive specifications of four proper subsets of Ω.

5. Let $\Omega = \{x|\, x$ is a word listed in *Webster's Third New International Dictionary*$\}$. What relations can you discern among the subsets $A = \{x|\, x \in \Omega$ and x contains at most one vowel$\}$, $B = \{x|\, x \in \Omega$ and x is a one-letter word$\}$, $C = \{x|\, x$

$\in \Omega$ and x contains exactly two vowels and three consonants), and $D = \{x | x$ is the word comet$\}$? What are the complements of A and of C?

6. Decide which of the following statements are true of all subsets of a set Ω, and devise counter examples to those that are not.
(**a**) A and A' are disjoint.
(**b**) If A and B are disjoint, then $B = A'$.
(**c**) If $A \subset B$, then $B' \subset A'$.
(**d**) If A and B are disjoint, then $A \subset B'$.
(**e**) If $A \subset B$, then A' and B are disjoint.
(**f**) If $A \subset B$, then A and B' are disjoint.
(**g**) If $A \subset B$ and if B and C are disjoint, then A and C are disjoint.
(**h**) If A and B are disjoint and B and C are disjoint, then A and C are disjoint.

7. Show that if $\Omega = \{\omega_1, \omega_2, \ldots, \omega_n\}$ then there are 2^n subsets of Ω (use induction on n).

8. Refer to the four possibilities of joint membership or nonmembership of an element x of Ω in subsets A and B of Ω. Describe, as fully as possible, the consequences of supposing that there is no x in Ω that satisfies
(**a**) Alternative (1) or alternative (2).
(**b**) Alternative (2) or alternative (4).
(**c**) Alternative (1) or alternative (2) or alternative (4).

1.4 Operations on sets

We continue in this section with a fixed universal set Ω. Two operations that are performed on the subsets of Ω are introduced here, each of which results in another subset of Ω. Some time will be spent in exploring these operations and their interconnections with the notions of inclusion, disjointness, and complementation.

Suppose that A and B are two subsets of Ω. The *union* of A and B, written $A \cup B$, is the subset of Ω consisting of elements x that are in A or in B or in both A and B. In other words, an element x of Ω is in $A \cup B$ if and only if x is in *at least one* of the two sets A, B. The *intersection* of A and B, written $A \cap B$, is the subset of Ω consisting of elements x that are in A and in B. Equivalently, an element x of Ω is in $A \cap B$ if and only if x is in *both* of the sets A and B. Note that the order in which A and B are written has no effect in these definitions. Thus, $A \cup B = B \cup A$ and $A \cap B = B \cap A$.

In a certain sense the operations of union and intersection can be viewed as enlarging and shrinking ones, respectively. In the case of $A \cup B$, we can argue as follows: every element x of Ω that is in A ($x \in A$)

must be in at least one of the two sets A and B ($x \in A \cup B$); thus, $A \subset A \cup B$. Of course, $B \subset A \cup B$ for the same reason. In the case of $A \cap B$, every element x of Ω that is in both A and B ($x \in A \cap B$) must be in the set A ($x \in A$). Therefore, $A \cap B \subset A$. Similarly, $A \cap B \subset B$.

If we assume various things about the subsets A and/or B of Ω, we may be able to find $A \cup B$ and $A \cap B$ explicitly. For example, let us suppose that $A \subset B$. Accordingly, every element x of Ω that is in A is also in B. When this is so, $x \in A$ implies $x \in A$ *and* $x \in B$, but then $x \in A \cap B$. Thus, $A \subset A \cap B$. We have already seen above that $A \cap B \subset A$ so it must be that $A = A \cap B$. Similarly, $x \in A \cup B$ implies that x is in at least one of the sets A and B, but then $x \in B$ (Why?). This argument gives the inclusion $A \cup B \subset B$ which, together with $B \subset A \cup B$, shows that $B = A \cup B$, if $A \subset B$. Inasmuch as we know that certain inclusions hold for any subset A of Ω, we can exhibit the following results of forming unions and intersections:

For any subset A of Ω,

$$\varnothing \cup A = A \qquad \Omega \cup A = \Omega \qquad A \cup A = A \tag{1.1}$$

$$\varnothing \cap A = \varnothing \qquad \Omega \cap A = A \qquad A \cap A = A$$

The disjointness of two sets A and B can be written directly in terms of their intersection as $A \cap B = \varnothing$. A set A and its complement are disjoint, and we find that

$$A \cup A' = \Omega \quad \text{and} \quad A \cap A' = \varnothing \quad \text{for any subset } A \text{ of } \Omega \tag{1.2}$$

Some more subtle facts can be ascertained about unions and intersections jointly. We give in Eq. (1.3) the *distributive law*.

$$\begin{gathered} A \cap (B \cup C) = (A \cap B) \cup (A \cap C) \\ \text{for any subsets } A, B, \text{ and } C \text{ of } \Omega \end{gathered} \tag{1.3}$$

[The use of parentheses here and later is the same as that in ordinary algebra, viz., to give the order in which operations are to be performed. In this instance, the left-hand side of Eq. (1.3) is the result of first forming the union of B and C and then intersecting this set with A; the right-hand side of Eq. (1.3) is the set that arises from first forming the intersection of A with B and A with C, then taking the union of the resulting sets.] To prove Eq. (1.3), we show that $A \cap (B \cup C) \subset (A \cap B) \cup (A \cap C)$ and that $(A \cap B) \cup (A \cap C) \subset A \cap (B \cup C)$. First we suppose that $x \in A \cap (B \cup C)$. Successively, we may argue that $x \in A \cap (B \cup C)$ implies $x \in A$ and $x \in B$ or $x \in C$ or both, which implies $x \in A$ and B, or $x \in A$ and C, or both, which implies $x \in (A \cap B) \cup (A \cap C)$. On the other hand, suppose $x \in (A \cap B) \cup (A \cap C)$. Then

$x \in A$ and B, or $x \in A$ and C, or both, which implies $x \in A$, and $x \in B$ or $x \in C$ or both, which implies $x \in A \cap (B \cup C)$. These inclusions give Eq. (1.3).

Next we give the pair of laws called the *De Morgan laws*:

$$(A \cup B)' = A' \cap B'$$

$$(A \cap B)' = A' \cup B' \tag{1.4}$$

for any subsets A and B of Ω

We prove the first of these with the method used above. We suppose that $x \in (A \cup B)'$. Successively, $x \in (A \cup B)'$ implies $x \notin A \cup B$, which implies $x \notin A$ and $x \notin B$, which implies $x \in A'$ and $x \in B'$, which implies $x \in A' \cap B'$. Conversely, $x \in A' \cap B'$ implies $x \in A'$ and $x \in B'$, which implies $x \notin A$ and $x \notin B$, which implies $x \notin (A \cup B)$, which implies $x \in (A \cup B)'$.

The operations of union and intersection are not at all restricted to two sets. If $A_1, A_2, \ldots, A_k$ are all subsets of Ω, the *union* of $A_1, A_2, \ldots, A_k$, written $A_1 \cup A_2 \cup \cdots \cup A_k$ or $\bigcup_{i=1}^{k} A_i$, is the subset of Ω that consists of all elements x that are in *at least one* of $A_1, A_2, \ldots, A_k$. The *intersection* of $A_1, A_2, \ldots, A_k$, written $A_1 \cap A_2 \cap \cdots \cap A_k$ or $\bigcap_{i=1}^{k} A_i$, is the subset of Ω that consists of the elements x that belong to *every one* of $A_1, A_2, \ldots, A_k$. Note that the order in which these sets are written has no bearing on the resulting subset and that if $k = 2$, these definitions agree with those given previously.

The analogues to Eqs. (1.3) and (1.4) hold for k subsets by the same kind of arguments. Equation (1.5) is the *distributive law* and Eq. (1.6) is the *De Morgan laws.*

$$A \cap \left(\bigcup_{i=1}^{k} A_i\right) = \bigcup_{i=1}^{k} (A \cap A_i) \quad \text{for any subsets } A, A_1, A_2, \ldots, A_k \text{ of } \Omega \tag{1.5}$$

$$\left(\bigcup_{i=1}^{k} A_i\right)' = \bigcap_{i=1}^{k} A_i'$$

$$\left(\bigcap_{i=1}^{k} A_i\right)' = \bigcup_{i=1}^{k} A_i' \quad \text{for any subsets } A_1, A_2, \ldots, A_k \text{ of } \Omega \tag{1.6}$$

Equation (1.5) gives the distributive law, and Eq. (1.6) gives the DeMorgan laws for subsets $A_1, A_2, \ldots, A_k$, where $k \geq 2$.

Subsets $A_1, A_2, \ldots, A_k$ having the property that $\bigcup_{i=1}^{k} A_i = \Omega$ are called *exhaustive*. This terminology reflects the fact that *every* element x of Ω is in at least one of the subsets $A_1, A_2, \ldots, A_k$. Subsets $A_1, A_2, \ldots, A_k$ for which $A_i \cap A_j = \varnothing$ for all choices of $i \neq j$ are called *mutually exclusive* or *mutually disjoint*. In this case, an element x of Ω is in at most one of the subsets $A_1, A_2, \ldots, A_k$. Combining these two properties, we find that subsets $A_1, A_2, \ldots, A_k$ of Ω that are both exhaustive and mutually exclusive are such that every element x of Ω is in *exactly* one of the sets $A_1, A_2, \ldots, A_k$. Exhaustive and mutually exclusive subsets play a special role in certain applications, as we shall show later. Note from Eq. (1.2) that an example of subsets with these two properties is A and A' for any subset A of Ω. For these special circumstances, we set out a translation of Eq. (1.5):

$$A = \bigcup_{i=1}^{k} (A \cap A_i) \quad \text{whenever } A \text{ is a subset of } \Omega \text{ and } A_1, A_2, \ldots, A_k \text{ are exhaustive and mutually exclusive subsets of } \Omega \tag{1.7}$$

Notice that the subsets $(A \cap A_1), (A \cap A_2), \ldots, (A \cap A_k)$ are mutually exclusive whenever $A_1, A_2, \ldots, A_k$ are mutually exclusive.

Example 1.2

Let $\Omega = \{x \mid x \text{ was a baby born in 1911}\}$, and consider the following subsets of Ω: $A_1 = \{x \mid x \text{ was born in a month of 28 days}\}$, $A_2 = \{x \mid x \text{ was born in a month of 30 days}\}$, $A_3 = \{x \mid x \text{ was born in a month of 31 days}\}$, $A_4 = \{x \mid x \text{ was a male baby}\}$, and $A_5 = \{x \mid x \text{ survives to the present}\}$. Clearly, A_1, A_2, and A_3 are exhaustive and mutually exclusive subsets of Ω. For A_4 playing the role of A in Eq. (1.7), the subset of male members of Ω is the union of the mutually exclusive subsets of males born in February, of males born in April, June, September, or November, and of males born in the remaining months. A_4 and A_4' are also exhaustive and mutually exclusive subsets of Ω. We could obtain from Eq. (1.7) that $A_5 = (A_5 \cap A_4) \cup (A_5 \cap A_4')$. In words, the set of surviving members of Ω is the union of the sets of surviving males and surviving females.

The first part of Eq. (1.4), applied to the sets A_4 and A_5, states that a member of Ω who is not "male or surviving or both" must be a nonsurviving female. The second part of Eq. (1.4) states, in this case, that a member of Ω who is not both male and surviving must be either female or nonsurviving or both.

EXERCISES 1.4

1. Let $\Omega = \{1, 2, 3, 4, 5, 6, 7, 8, 9, 10\}$, and consider the following subsets of Ω: $A = \{3, 6, 9\}$, $B = \{2, 3, 5, 7\}$, $C = \{1, 3, 5, 7, 9\}$, and $D = \{1, 2, 3, 4, 5, 6, 7\}$. List each of the following subsets of Ω:

(a) $A \cap B$.
(b) $A \cup B$.
(c) $A \cup B \cup C$.
(d) $A \cap B \cap C$.
(e) $A \cap (B \cup C)$.
(f) $A \cap (B \cup C \cup D)$.
(g) $(C \cap D)'$.
(h) $(A \cup C \cup D)'$.
(i) $B \cup D$.
(j) $(A' \cap (B \cup C)) \cap (A \cup D)$.

2. Let Ω be the set of undergraduate students at a university, and let A_1, A_2, A_3, and A_4 be, respectively, the subsets of freshman, sophomore, junior, and senior students. Further, let B be the subset of female students, and let C be the subset of students owning cars. For each of the following sets, give an alternative verbal specification.
(a) $A_1 \cup A_2$.
(b) $C \cap (A_1 \cup A_2)$.
(c) $B \cap A_3$.
(d) $(C \cup B)'$.
(e) $B \cap (A_1 \cup A_2 \cup A_3)'$.
(f) $(B \cap (A_1 \cup A_2 \cup A_3))'$.

3. Consider the sets given in Exercise 2. What is the meaning of each of the following conditions?
(a) $B = \varnothing$.
(b) $B \cap C = \varnothing$.
(c) $B \cup C = \varnothing$.
(d) $A_4 \subset C$.
(e) $C \subset A_3 \cap A_4$.
(f) $A_1 \cup C = \Omega$.
(g) $((A_1 \cup A_2) \cap C)' = B$.

4. Determine which of the following statements are true of all subsets of a set Ω, and devise counter examples to those that are not.
(a) $(A \cup B')' = A' \cap B$.
(b) If $A \cup B = \Omega$ and $A \cap B = \varnothing$, then $A = B'$.
(c) If $A \cup B = \Omega$, then $A \subset B'$.
(d) If $A \cup B = \Omega$, then $A' \subset B$.
(e) If $A \subset B'$, then $A \cup B = \Omega$.
(f) $((A \cap B)')' = A \cap B$.
(g) $(A \cap B)' \subset A$.
(h) $(A \cap B) \cup C = A \cap (B \cup C)$.

5. Show that $\bigcap_{i=1}^{k} A_i \subset A_1$ and that $A_1 \subset \bigcup_{i=1}^{k} A_i$.

6. Show that if $A_1, A_2, \ldots, A_k$ are exhaustive subsets of Ω then so are $A_1, A_2, \ldots, A_k, A_{k+1}$.

7. Show that if $A_1, A_2, \ldots, A_k$ are mutually exclusive subsets of Ω then so are $A_1, A_2, \ldots, A_{k-1}$.

8. Prove the second part of Eq. (1.4).

9. Prove the second part of Eq. (1.6).

10. Show that if A, B, and C are subsets of Ω then a second distributive law holds, viz., $A \cup (B \cap C) = (A \cup B) \cap (A \cup C)$.

11. Prove the distributive law of Exercise 10 by assuming Eqs. (1.3) and (1.4).

12. Let $N(A)$ be the number of elements in A for each subset A of Ω. State each of the following conditions in terms of these numbers.
    (a) $A = \emptyset$.
    (b) A and B are disjoint.
    (c) A is a proper subset of Ω.
    (d) $A \subset B$.
    (e) $A = B$.

13. Let $N(A)$ be as given in Exercise 12. Refer to the four possibilities listed in Section 1.3 for joint membership or nonmembership of an element x in subsets A and B of Ω. Suppose that there are 14 elements satisfying alternative (1), 12 satisfying alternative (2), 8 satisfying alternative (3), and 15 satisfying alternative (4). With this information, determine $N(\Omega)$, $N(A)$, $N(A \cap B)$, $N(A \cup B)$, $N(A')$, $N(B')$, and $N(A' \cup B')$.

14. The *symmetric difference* of two subsets A and B of Ω, written $A \Delta B$, is given by $A \Delta B = (A \cap B') \cup (A' \cap B)$. What subsets of Ω are given by $A \Delta \emptyset$, $A \Delta \Omega$, $A \Delta A$, and $A \Delta A'$?

1.5 Product sets

In this section we consider universal sets having a special structure and some of the attending subsets. In order to do this, we must first bring out certain basic matters. We temporarily abandon the restrictions that all objects x are members of a fixed universal set Ω and that all sets A are subsets of Ω.

A pair of objects x and y of which x is considered to be first and y to be second is called an *ordered pair* and is written (x, y). An ordered pair (x, y) is viewed as a new object and as such is distinct from (y, x) for, in the latter case, y is considered to be first. Imagine, for example, the ticket presented by the Republican party for national election. Such a ticket is composed of a candidate x for president and a candidate y for vice-president. This ticket may be represented by the ordered pair (x, y) and then should be distinguished from that represented by (y, x).

Let A and B be two sets. The *Cartesian product* of A and B, written $A \times B$, is the set of all ordered pairs (x, y) for which $x \in A$ and $y \in B$. Notice the ordering implied here: the first objects are members of the set given on the left. Generally, we find that the set $A \times B$ is not equal to the set $B \times A$ for just this reason.

When A and B are finite sets, say $A = \{x_1, x_2, \ldots, x_m\}$ and $B = \{y_1, y_2, \ldots, y_n\}$, a convenient tabling of the ordered pairs in $A \times B$ can be done as in Figure 1.3. Thus, $A \times B$ is the set of ordered pairs given in the table, of which there are $m \cdot n$.

$A \backslash B$	y_1	y_2	...	y_n
x_1	(x_1, y_1)	(x_1, y_2)	...	(x_1, y_n)
x_2	(x_2, y_1)	(x_2, y_2)	...	(x_2, y_n)
.	.	.		.
.	.	.		.
.	.	.		.
x_m	(x_m, y_1)	(x_m, y_2)	...	(x_m, y_n)

FIGURE 1.3

$A \backslash B$	Clay	Day	Emerson
Adams	(Adams, Clay)	(Adams, Day)	(Adams, Emerson)
Brown	(Brown, Clay)	(Brown, Day)	(Brown, Emerson)

FIGURE 1.4

If the Republican party regards Adams and Brown as presidential timber and Clay, Day, and Emerson as vice-presidential material, then the Cartesian product of the sets $A = \{\text{Adams, Brown}\}$ and $B = \{\text{Clay, Day, Emerson}\}$ will provide possible election tickets. In tabled form, these are as shown in Figure 1.4.

The set $A \times B$ in Figure 1.4 contains $2 \cdot 3 = 6$ elements in all and consequently $2^6 = 64$ subsets. Let us compare the following three subsets: $A_1 = \{(\text{Adams, Clay}), (\text{Brown, Day})\}$, $A_2 = \{(\text{Adams, Clay}), (\text{Adams, Day})\}$, and $A_3 = \{(\text{Adams, Clay}), (\text{Brown, Clay})\}$. A_2 and A_3 are themselves Cartesian product sets. In particular, $A_2 = \{\text{Adams}\} \times \{\text{Clay, Day}\}$ and $A_3 = \{\text{Adams, Brown}\} \times \{\text{Clay}\} = A \times \{\text{Clay}\}$. On the other hand, A_1 is not a product set. A_3 in the present context could be interpreted as the set of tickets nominating Clay for vice-president.

For the general product set $A \times B$ as given in Figure 1.3, these same types of subsets may be found. For example, $A_1 = \{(x_1, y_1), (x_2, y_1), (x_1, y_2)\}$ is not a product set for, with x_2 as a first entry and y_2 as a second, the element (x_2, y_2) would be required. In fact, $A_2 = \{(x_1, y_1), (x_2, y_1), (x_1, y_2), (x_2, y_2)\} = \{x_1, x_2\} \times \{y_1, y_2\}$. The subset $A_3 = A \times \{y_1, y_2\}$—i.e., that which consists of all ordered pairs in the first two columns of Figure 1.3—is analogous to the set A_3 of the previous paragraph.

An r-tuple of objects $x_1, x_2, \ldots, x_r$ of which x_1 is considered to be first, x_2 to be second, ..., x_r to be rth is called an *ordered r-tuple* and is written

$(x_1, x_2, \ldots, x_r)$. These new objects are the basis of forming new sets in a manner analogous to what has been done with $r = 2$ above.

Let $A_1, A_2, \ldots, A_r$ be sets. The *Cartesian product* of $A_1, A_2, \ldots, A_r$, written $A_1 \times A_2 \times \cdots \times A_r$ or $\displaystyle\mathop{\times}_{i=1}^{r} A_i$, is the set of all ordered r-tuples $(x_1, x_2, \ldots, x_r)$ for which $x_1 \in A_1$, $x_2 \in A_2$, $\ldots$, $x_r \in A_r$. Once again the order in which the sets $A_1, A_2, \ldots, A_r$ are written is essential.

A tabling of the ordered r-tuples of $A_1 \times A_2 \times \cdots \times A_r$ comparable to that of Figure 1.3 requires a good deal of patience, for $r > 2$. Correspondingly, the insight that it provides decreases very rapidly as r gets

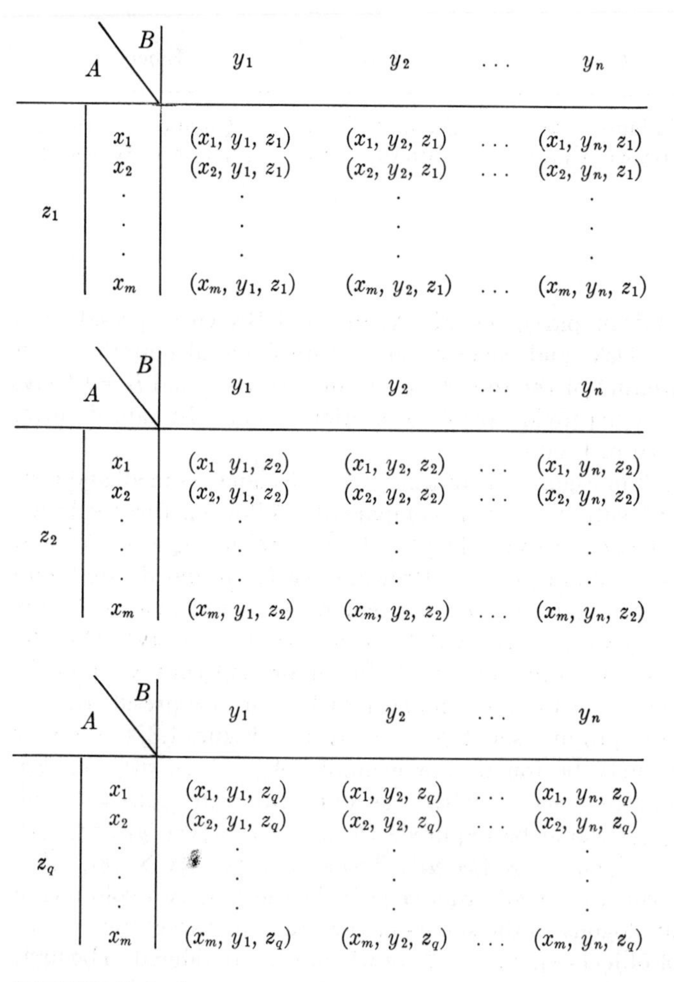

FIGURE 1.5

larger. We persist, nonetheless, to $r = 3$. Let us suppose that $A = \{x_1, x_2, \ldots, x_m\}$, $B = \{y_1, y_2, \ldots, y_n\}$, and $C = \{z_1, z_2, \ldots, z_q\}$ and consider $A \times B \times C$. For each member of C, we give a table of ordered 3-tuples (or triples) with that member as the third object (Figure 1.5).

There are q subtables, each of which has $m \times n$ ordered triples; hence there are $m \times n \times q$ ordered triples in $A \times B \times C$.

There are many product subsets of $A \times B \times C$. A simple subset of $A \times B \times C$ is itself a Cartesian product set; for example, $\{(x_1, y_1, z_1)\} = \{x_1\} \times \{y_1\} \times \{z_1\}$. Another is the set of all ordered triples from the first subtable, which may be written in the form $A \times B \times \{z_1\}$. Yet another is $A \times \{y_1, y_2\} \times \{z_1\}$, the set of all ordered triples of the first and second columns of the first table. On the other hand, $\{(x_1, y_1, z_1), (x_1, y_2, z_2)\}$ is not a product subset of $A \times B \times C$.

One case of general Cartesian product sets deserves a separate mention. It is the case in which we consider $A_1 \times A_2 \times \cdots \times A_r$, where each A_i is the same set, say A. We will refer to this as the Cartesian product of A with itself r times, for which we will write A^r for the set of r-tuples. There is nothing new in this; for example, if $A = \{1, 2\}$, then $A^2 = A \times A = \{(1, 1), (1, 2), (2, 1), (2, 2)\}$ and $A^3 = A \times A \times A = \{(1, 1, 1), (1, 1, 2), (1, 2, 1), (2, 1, 1), (1, 2, 2), (2, 1, 2), (2, 2, 1), (2, 2, 2)\}$.

Before concluding, we point out that throughout the present section certain Cartesian product sets have been (implicitly) regarded as temporary universal sets. In future chapters, the natural universal set may turn out to be such a product set and we shall wish to exploit the fact. When this is so, say $\Omega = \bigtimes_{i=1}^{r} \Omega_i$, the discussion will be limited to the elements of Ω (which are ordered r-tuples) and to the subsets of Ω.

EXERCISES 1.5

1. Which of the following are Cartesian product sets? Write them in product form.

(**a**) $\{(1, 2), (2, 2)\}$. (**b**) $\{(2, 2), (1, 2), (2, 1)\}$.
(**c**) $\{(2, 2), (2, 1), (2, 3)\}$. (**d**) $\{(2, 2), (1, 2), (1, 3), (2, 3)\}$.
(**e**) $\{(1, 2, 3), (2, 2, 3), (2, 1, 3)\}$. (**f**) $\{(1, 2, 3, 4, 5, 6), (1, 1, 3, 4, 4, 6)\}$.
(**g**) $\{(1, 2), (1, 2, 3)\}$. (**h**) $\{(1, 2, 3, \text{go})\}$.

2. Specify $A \times B$ by listing members when
(**a**) $A = \{\text{stop, go}\}$, $B = \{\text{male, female}\}$.
(**b**) $A = \{\text{stop, go}\}$, $B = \{3, 1, 2, 4\}$.
(**c**) $A = \{\text{stop, go}\}$, $B = \{(1, 2), (2, 2)\}$.

3. Specify $A \times B \times C$ by listing members when $A = \{\text{stop, go}\}$, $B = \{1\}$, and $C = \{39.5, 83.4, 27.6\}$.

4. Which of the following statements are true?
   (a) $\varnothing \times B = \varnothing$.
   (b) If $A \subset C$, then $A \times B \subset C \times B$.
   (c) If $A = C \times D$ and $B = E \times F$, then $A \cap B = (C \cap E) \times (D \cap F)$.
   (d) $A \times B \subset A \times B \times C$.

5. Prove or disprove: $A \times B = B \times A$ implies $A = B$.

6. Prove or disprove: $A^k \times A^r = A^{k+r}$.

7. Let $\Omega = A \times B$, where $A \times B$ is as shown in Figure 1.3. Let C and D be proper subsets of Ω which are themselves Cartesian product sets. Is C' a product set? Is $C \cup D$ or $C \cap D$ a product set?

8. Let $\Omega = A \times B$, where $A \times B$ is as shown in Figure 1.3. Find the number of proper subsets of Ω that are themselves product sets.

CHAPTER TWO

Sample Spaces

2.1 Introduction

In this chapter we establish the mathematical context in which we will study the theory of probability. Before we can adopt this mathematical context, however, we must spend some time clarifying the relationship of the theory we will study to the situations that allow its application and that have historically provided the impetus to the theoretical material.

The two mathematical concepts introduced here are that of a sample space (Section 2.3) and that of the events of a sample space (Section 2.4). These concepts are, respectively, a universal set and its subsets called by new names because of the particular setting. We have attempted to move carefully from the real-life situation to the mathematical context of a sample space and its events with the expectation that subsequent chapters may be primarily concerned with mathematics. You should then refer to this chapter for answers to your questions on the relevance of theory to practice.

2.2 Probability theory

The notion of chance or probability has several common uses. We all are familiar with the use illustrated by such statements as, "He was probably a contented man," or "That would probably break her heart," or "The cosmological theory of the Big Bang is probably false," or "Had Napoleon lived longer, he would probably have attempted a return to power," and so on. The notion of probability that is used in these statements may be described roughly as an impressionistic rendering of the speaker's state of mind—an indication of the speaker's judgments about various things.

However, most of us are familiar with another notion of chance or probability, distinct from the one illustrated above. This one might be called physical or statistical probability. It is illustrated by statements of the following sort: "If a coin is tossed repeatedly, a head will probably turn up eventually," or "John is such a good marksman that if he were to shoot at a target at 50 yards he would probably not miss," or "The chances of a case of tuberculosis being fatal are vastly smaller today than they were 50 years ago," or "Heavy smoking probably means high susceptibility to lung cancer," and so on.

A little reflection shows that there is an important qualitative difference between the statements of the first paragraph and those of the second. Namely, while the former are judgments made by the speaker and cannot be subjected to any practicable test of their validity, the latter relate to the observation of the outcome of a possible experiment, and, if we so chose, we could try to convince ourselves of the validity of these statements by performing the said experiments. Thus, the four statements of the second paragraph could be tested by performing, say, the following four experiments: tossing a coin repeatedly, having John shoot at a target at 50 yards, examining the records of fatalities from tuberculosis, observing the incidence of lung cancer among heavy smokers and comparing it with the incidence of the same disease among those who are not heavy smokers. Of course, if we were to actually try to describe a procedure for testing the validity of such statements, we would have to be both more precise and more subtle. For the present, however, we are not interested in describing *how* we may establish the validity or invalidity of the statements in the second paragraph. We are rather more interested in stressing that we *can* conceive of a procedure for doing so. On the other hand, it is not easy to conceive of such procedures for the statements in the first paragraph.

In this book, statements akin to those of the first paragraph will not concern us at all. Instead, we shall be concerned with the notion of physical or statistical probability. As we have illustrated above, such a notion will have as a reference point the outcomes of an underlying conceptual experiment. Indeed, we shall soon see that we will be interested in making meaningful statements that are even more precise than those of the second paragraph above.

Before we can talk of probabilities with any precision, we must agree upon the experiment and the outcomes of the experiment to which our statements are going to refer. Only when that has been decided will we have a frame of reference with respect to which our statements could have a satisfactory meaning. In the next few sections, we shall be setting up just such a frame of reference. At the present stage, however, it may not be out of place to spend some time discursively in clarifying the role of the theoretical material to follow in the next few chapters.

Mathematical disciplines often grow out of a need and a desire to understand and describe some of the formal properties of a physical phenomenon. The most familiar example of this is geometry. The classical motivation for the study of geometry is the desire to understand the formal properties of space, especially the properties of the relative location of objects in space.

It was realized during the nineteenth century that geometry as a discipline was concerned solely with the study of *relations* between undefined objects. Thus, geometers spend no time trying to analyze the "true nature" of a point or a line. What matters to them is that relations of a definite kind exist between "points" and "lines," e.g., two distinct points determine a line and two distinct lines in general determine a point. These relations between points and lines, along with some others, form the rules by which the game of geometry is played. There is nothing sacred about the rules; if they are changed we get a different game. Indeed, mathematicians have studied so-called non-Euclidean geometries whose rules are different from those postulated by Euclid. Now, our perception of space in the small is in excellent agreement with the postulates of Euclidean geometry, and, for many activities in life for which a geometrical model of space is desired, we single out Euclidean geometry as the one that is most useful. For example, in such engineering problems as surveying, the assumption is tacitly made that the geometrical relations between objects in space are the same as those described by Euclidean geometry. However, this does not make Euclidean geometry intrinsically distinguished among all possible geometries. They are all horses of the same breed, and the fact that the Euclidean horse is worked more than the others does not make it any more (or less) a horse.

Thus, from a philosophical standpoint, we can make a distinction between the subject matter of geometry as a discipline and the application of a particular kind of geometry to problems in real life. The former is a mathematical study. The latter is, we might say, a physical endeavor. Of course, given a situation that calls for the application of a geometrical theory and a variety of geometrical theories from which we are asked to choose one that best fits the situation, it may be extremely hard to decide which one fits best. (Modern physics has been beset with precisely such a choice in some of its theories.) As a further complication, the alternative geometries available may be equally capable of explaining the grosser aspects of the physical situation, so that our choice must be guided by intuition. It should be clear that in such cases there are subtle problems that cannot be decided on mathematical grounds alone. Thus, while the study of geometry is indeed a mathematical endeavor, our success in applying the mathematical theory to real-life problems depends upon the keenness of our intuition.

Similar comments may be made about other mathematical theories. Axiomatically, mathematics consists solely of the study of *relations* between objects and collections of objects. The "true nature" of the objects is of no relevance to the formal structure of mathematics. On the other hand, it can and often does happen that relations that are observed to exist between certain objects in a situation in real life may be adequately described by means of one or more mathematical theories. Here one of several alternative mathematical models may have to be selected. The choice may be difficult and may force a heavy reliance on our intuition.

In the course of our study, there will be ample opportunity to observe how the above comments apply to the mathematical theory of probability. In anticipation, we may describe the scope of the theory roughly as follows. We fix a conceptual experiment and suppose we have a reasonable description of the outcomes of such an experiment. Various events may occur when the experiment is performed. We shall be interested in obtaining precise descriptions of all events that may occur and in assigning to each event a numerical measure that we shall call its *probability*. The mathematical theory may be said to consist of the study of the relations between the various events and their probabilities. Thus, the *mathematical* study begins after the possible events have been agreed upon and their probabilities have been specified as mathematical objects. Of course, if the experiment being contemplated is an idealization of a real-life situation, then, in order that our mathematical theory be useful for making predictions, our assignment of probabilities to various events must have a sound basis, for different assignments of probabilities will result in different probabilistic models for the same experiment, and we are thus faced with the kind of choice discussed in the paragraphs above.

There are then two distinct aspects to successfully applying probability theory to practical problems. Firstly, there is the formal aspect of studying the mathematical theories that may conceivably be of use. Secondly, there is the more intangible aspect of learning to single out a given mathematical theory as being the one that has the best promise of success in a given situation. If probability theory is regarded as a purely mathematical discipline, then only the first aspect is important. If, on the other hand, we are to enjoy the full flavor of the theory, the second aspect is indispensable. It is here that we are thrown back on intuition and experience.

The formal material of the following chapters is devoted to developing some of the mathematical aspects of probability theory. However, this is not because we feel that the other aspect is not worthy of study. Rather, it is because we feel that the other aspect can only be learned by illustration and by practice. We shall therefore rely on illustrative examples to bridge the gap between theory and applications. Bear in mind therefore that the exercises are an integral part of the material to be studied.

2.3 The sample space

The mathematical theory of probability that we shall discuss will bear upon real-life situations only in the context of an experiment, either real or conceptual. You have previously encountered probability statements referring to outcomes of such experiments as tossing a coin a hundred times, throwing two dice, dealing a bridge hand from a well-shuffled deck of cards, participating in the Irish Sweepstakes, selecting a random sample of wage-earners and observing the length of their working day, observing the life span of various people, observing the size of flowers produced by cross pollination of two different varieties of a plant, observing the number of defectively filled bottles in a bottling plane, and observing the frequency of air crashes. But before we can talk about the probabilities of various events that may occur when such experiments are performed, we must agree on the universe of outcomes that we are interested in studying. In other words, we must agree on the outcomes of the experiment that will be regarded as possible. The classic illustration of what we mean here is the following: when a coin is tossed, it may fall neither heads nor tails but may stand on edge. In trying to describe possible outcomes of the experiment of tossing a coin, we may legitimately think of three possible outcomes: heads, tails, or stand on edge. However, the third outcome in this list is rather artificially conceived. It is legitimate from the logical or philosophical standpoint, but if we build a theory ruling out this outcome, the applicability of such a theory could hardly be less than that of a theory that considers it a possible outcome.

By agreeing on the universe of possible outcomes, we are of course resorting to idealization. Such idealization is common to all branches of science, and it provides us with the economy that makes feasible the development of a theory. In choosing the universe of possible outcomes, we are guided by convenience insofar as this convenience does not compromise any essential aspects of the experiment.

For example, we might choose and describe the possible outcomes of the experiment of tossing a coin thrice in the following way. First, we agree that each toss must result either in heads or tails. One possible outcome of the experiment is then that all tosses result in heads. Using the letter H for heads, we may describe this outcome by the symbol HHH. Similarly, the symbol HTT would describe the outcome that the first toss resulted in heads and the second and third in tails. Using this symbolism, we find that eight outcomes are possible: HHH, HTH, HHT, HTT, THH, TTH, THT, TTT. Notice here that we resort to a shorthand description of the outcomes when a verbal description would require somewhat more space, i.e., we label the outcomes by means of symbols. We could even go a step further. In the above case, we could establish the following shorthand. We let ω_1 stand for the symbol HHH, ω_2 for HTH, ω_3 for HHT, ω_4 for HTT, ω_5 for THH, ω_6 for TTH, ω_7 for THT, and ω_8 for TTT. Now we may describe the universe of possible outcomes even more succintly by the list $\omega_1, \omega_2, \omega_3, \omega_4, \omega_5, \omega_6, \omega_7, \omega_8$. Of course, this list has contact with our experiment only because we have a "dictionary" that identifies $\omega_1, \omega_2, \ldots, \omega_8$, respectively, with the possible outcomes.

For another example, let us consider the experiment of tossing a die twice. First, we can agree that the die will land with one of the six faces uppermost. One possible outcome is, for example, that the face marked with one dot is uppermost on both throws. We may represent this by the symbol 11. Similarly, 31 could represent the outcome that the first throw resulted in a score of 3 while the second resulted in a score of 1. Using such a convention, we determine that there are 36 possible outcomes (Figure 2.1). Again, if we chose, we could carry the labelling a step further and name the outcomes in the first column $\omega_1, \omega_2, \ldots, \omega_6$; those in the second column $\omega_7, \omega_8, \ldots, \omega_{12}$; and so on through ω_{36}. Then we may list the possible outcomes simply as $\omega_1, \omega_2, \ldots, \omega_{36}$. Once again, this list has contact with our experiment only because of the "dictionary" that identifies members of this list with possible outcomes.

As another example, consider the following. As a part of a study designed to gain information about family size and composition, a newly married couple is observed throughout their productive life, and a record is kept of the total number of their male and female offspring. First, we

11	21	31	41	51	61
12	22	32	42	52	62
13	23	33	43	53	63
14	24	34	44	54	64
15	25	35	45	55	65
16	26	36	46	56	66

FIGURE 2.1

might make a preliminary agreement that the couple will produce no more than, say, 30 offspring. Then one possible outcome of the experiment would be that there are 5 male and 3 female offspring. We might represent this outcome by the ordered pair (5, 3). Similarly, (0, 7) could represent the outcome of no male and 7 female offspring. We arrive generally at possible outcomes represented by an ordered pair of integers (n_1, n_2), where (n_1, n_2) describes the outcome n_1 male and n_2 female offspring. Of course, $n_1 \geq 0$ and $n_2 \geq 0$, and we have agreed that $n_1 + n_2 \leq 30$. Thus, in this example, all possible outcomes can be represented by all ordered pairs (n_1, n_2) of nonnegative integers whose sum is between 0 and 30 inclusive. Now, if we chose, we could list all possible outcomes with this representation. It would be rather large for there are 496 such ordered pairs. We could also, as before, label the different possible outcomes as $\omega_1, \omega_2, \ldots, \omega_{992}$ in some order. Here again, the labels define a "dictionary" by which each particular possible outcome may be identified. As in the previous examples, we are led to represent the totality of possible outcomes by a collection of symbols and to identifying precisely each symbol with a possible outcome of the experiment.

In building a mathematical theory, one tries to isolate for study the features of the subject matter that are general, i.e., those that are shared by diverse particular situations. In all the illustrations above, the universe of possible outcomes was described by a *collection* of symbols. In contrast, the dictionary that identifies each symbol as a possible outcome necessarily varies from one case to the next. Thus, for our purposes, the universe of possible outcomes of an experiment may be thought of as a collection of symbols. In other words, the mathematical concept that we will utilize for representing the totality of possible outcomes of an experiment is the concept of a *set.*

We fix a set Ω once and for all, and its elements will be thought of as possible outcomes of a conceptual experiment. The set Ω we will call the *sample space,* and its elements we will call *sample points.* The sample points may be denoted by $\omega_1, \omega_2, \ldots$, etc. *In most of this book, we shall only be concerned with finite sample spaces,* i.e., *we shall be considering experiments with only a finite number of possible outcomes.*

Considered as a mathematical object, the sample space is no more than a set. Why then must it be given a new name? The reason is twofold. On the one hand, by referring to the given set Ω as the sample space, we are recalling that in applications of the theory, the elements of Ω are to be thought of as outcomes of some experiment. On the other hand, the term "sample space" originated in statisticians' usage, was subsequently borrowed by probabilists, and is now firmly established.

Given an experiment, it requires considerable skill to decide on a set Ω that will serve as an adequate sample space. Indeed, to a certain extent

ω_1: 11	ω_7: 21	ω_{13}: 31	ω_{19}: 41	ω_{25}: 51	ω_{31}: 61
ω_2: 12	ω_8: 22	ω_{14}: 32	ω_{20}: 42	ω_{26}: 52	ω_{32}: 62
ω_3: 13	ω_9: 23	ω_{15}: 33	ω_{21}: 43	ω_{27}: 53	ω_{33}: 63
ω_4: 14	ω_{10}: 24	ω_{16}: 34	ω_{22}: 44	ω_{28}: 54	ω_{34}: 64
ω_5: 15	ω_{11}: 25	ω_{17}: 35	ω_{23}: 45	ω_{29}: 55	ω_{35}: 65
ω_6: 16	ω_{12}: 26	ω_{18}: 36	ω_{24}: 46	ω_{30}: 56	ω_{36}: 66

FIGURE 2.2

the choice of Ω may even depend on what we decide to regard as an outcome. Let us consider a simple example. In the experiment of throwing a die twice, we saw that we could consider the 36 outcomes in Figure 2.2. Thus, we could choose $\Omega = \{\omega_1, \omega_2, \ldots, \omega_{36}\}$ as our sample space. Given this sample space, we can easily answer such questions as, "Which outcomes will result in a total score greater than 5?" or "Which outcomes have the property that the score on the second toss exceeds the score on the first toss by 2 or more?" or "Which outcomes result in an odd score on the second toss?" and so on.

Now suppose that we decide to ignore all other aspects of the outcome except the total score. We might then decide that the following outcomes are possible: the score is 2, the score is 3, . . . , the score is 12. There are 11 such outcomes, and a suitable sample space is $\tilde{\Omega} = \{\tilde{\omega}_1, \tilde{\omega}_2, \ldots, \tilde{\omega}_{11}\}$ where

$\tilde{\omega}_1$: the score is 2
$\tilde{\omega}_2$: the score is 3
.
.
.
$\tilde{\omega}_{11}$: the score is 12

We can now ask which of the sample spaces, Ω or $\tilde{\Omega}$, is appropriate for the experiment of tossing a die twice. To answer this question, we note that with the sample space $\tilde{\Omega}$ it is possible to answer such questions as, "Which outcomes in $\tilde{\Omega}$ result in a score bigger than 5?" or "Which outcomes in $\tilde{\Omega}$ result in an odd score?" However, using $\tilde{\Omega}$ as a description of possible outcomes of the experiment, we lose some information. For example, a question like "Which outcomes have the property that the score on the second throw is greater than the score on the first?" can easily be answered on the basis of Ω, but it cannot be answered at all on the basis of $\tilde{\Omega}$. The reason is, of course, that in concocting $\tilde{\Omega}$ we have ignored all aspects of the outcome except the total score. Thus, $\tilde{\Omega}$ only describes the universe of possible total scores. $\tilde{\Omega}$ is wholly inadequate for the studies that require a more detailed knowledge of the outcome. Now, of course, if the questions of interest involve only total score, then $\tilde{\Omega}$ is an adequate sample space. Often it is hard to decide in advance exactly what questions we will want to study. Thus, we should choose a sample space that is, roughly speaking, adequate to answer all possible questions about outcomes of the experiment. In the example above, Ω is such a sample space while $\tilde{\Omega}$ is not. In general, the choice of an appropriate sample space requires skill and experience.

EXERCISES 2.3

In each of Exercises 1 to 22, an experiment is described. Consider the totality of possible outcomes, and devise a sample space to represent these outcomes. Describe the nature of the elements of the sample space, and, given a sample point, show how to decide which outcome it represents.

1. A coin is tossed once.

2. An urn contains two balls, one black and one red. A ball is drawn from the urn, and its color is noted.

3. A machine manufactures a certain item. An item produced by the machine is tested to determine whether or not it is defective.

4. A die is thrown once.

5. An urn contains six balls, each colored differently from any of the others. A ball is drawn from the urn, and its color is noted.

6. A person is asked to which of the following categories he belongs: unemployed, self-employed, employed by the U.S. Government, employed by a state or local government, employed by a private organization, employed by an employer who does not fulfill any of the above descriptions. The person's answer is recorded.

7. A coin is tossed eight times.

8. A coin is tossed n times, where n is a given positive integer.

9. A die is tossed four times.

10. A die is tossed n times, where n is a given positive integer.

11. An urn contains ten chips numbered 1 through 10. A chip is drawn, its number is noted, and the chip is replaced. Another chip is drawn, and its number is noted.

12. The experiment of Exercise 11 is repeated, with the exception that the first chip is not replaced prior to drawing the second.

13. A winner and a runner-up for a beauty contest are selected from among ten beautiful girls named Anna, Betsy, Carol, Dolores, Eva, Franny, Gita, Hermione, Inge, and Joan.

14. First, second, and third prize winners are selected from among the bevy of Exercise 13.

15. An urn contains n chips numbered 1 through n. k chips ($k \leq n$) are drawn in succession, never being replaced, and the number on each chip is noted in order of its appearance.

16. A coin is tossed. If the coin falls heads, a die is thrown. If the coin falls tails, it is tossed twice more.

17. In Ruretania, a boy takes an examination at age 11. If he passes, he goes to a grammar school. If he fails, he goes to a technical school. A grammar school graduate takes the university entrance examination and receives one of three grades. If he receives the top grade, he may go to the University of Oxbridge; if he receives the second grade, he may go to the University of Redbrick; if he receives the bottom grade, he may enter politics or business. On the other hand, a technical school graduate takes a polytechnic institute entrance examination and receives one of three grades. If he receives the top grade, he may enter the institute; if he receives the second grade, he may take a job in a trade; if he receives the bottom grade, he enters politics. The experiment is that of observing a boy's career from age 10 through the above choices.

18. An engineer is in charge of quality control in a plant manufacturing a certain type of nut. He observes 500 nuts in succession as they are stamped out by a machine and notes whether each one is defective or nondefective.

19. A certain variety of plant can be one of four distinct genetic types. One thousand plants are observed, and the genetic type of each is determined and noted.

20. A disease is manifested in a cellular disorder in the blood cells of the patient. Eight distinct types of disorders have been found to occur with the disease. Blood samples are taken from 100 patients, and, in each case, the type of cellular disorder that has occurred with the disease is noted.

21. A company selects a president, a vice-president in charge of sales, and a vice-president in charge of operations from among eight candidates.

22. A traveling salesman must visit each of 15 cities. He makes an itinerary detailing his route.

23. In Exercises 1 to 22, point out any basic similarities between an exercise and the exercises that precede it. Between exercises of a similar nature, what differences are there in the answers?

24. Invent two experiments with the same features as the one given in Exercise 18. Do the same for Exercises 19, 21, and 22.

25. What is the relation of the sample space of Exercise 1 to that of Exercise 7? In the same light, consider Exercises 4 and 9.

26. Do Exercise 11 if there are n chips in the urn and the number of draws is k.

2.4 Events

Suppose a finite set Ω has been chosen as a sample space for a given conceptual experiment. The aim of this section is to describe a correspondence between subsets of Ω on the one hand and various events that may occur in the conceptual experiment on the other.

The intuitive notion of an "event" is best broached by illustration. For example, we speak of the event that when a coin is tossed thrice, heads will turn up twice or more. Now if a coin were tossed thrice and we are told that the outcome was HTT, say, we can tell whether or not the event "Heads twice or more" has occurred. Indeed, if the outcome were HTT, we would say that the event "Heads twice or more" has *not* occurred. Similarly, given any one of the possible outcomes, we are able to say whether or not this event occurred. For example, in the present instance, we would say that the event "Heads twice or more" has occurred if the outcome of the experiment is HHH, HTH, HHT, or THH. By the same token, if the outcome were not one of these four, we would say the event did not occur. Thus, in this case, the description "In three tosses of a coin, heads twice or more" enables us to classify the set of possible outcomes into two subsets, one subset consisting of the outcomes that imply the occurrence of the event and a second subset consisting of the remaining outcomes that do not imply this occurrence.

Let us take another example in a similar context, the experiment of tossing a coin four times. As in the case of three tosses, we might arrive at the descriptions of possible outcomes listed in Figure 2.3. Now we consider the event "In four tosses of a coin, the number of heads is greater than or equal to the number of tails." If the experiment results in, say, the outcome ω_6: HTHT, we would say that this event has occurred. Indeed, we see that the prescribed event will occur if the experiment results in ω_1, ω_2, ω_3, ω_4, ω_5, ω_6, ω_7, ω_9, ω_{10}, ω_{11}, or ω_{12}. On the other hand, if the experiment results in ω_8, ω_{13}, ω_{14}, ω_{15}, or ω_{16}, then the event has not occurred. Again, we have been able to classify the set of possible out-

ω_1: HHHH	ω_9: THHH
ω_2: HTHH	ω_{10}: TTHH
ω_3: HHTH	ω_{11}: THTH
ω_4: HHHT	ω_{12}: THHT
ω_5: HTTH	ω_{13}: TTTH
ω_6: HTHT	ω_{14}: TTHT
ω_7: HHTT	ω_{15}: THTT
ω_8: HTTT	ω_{16}: TTTT

FIGURE 2.3

comes into two subsets, one subset consisting of the outcomes that imply the occurrence of the event and the other subset consisting of the remaining outcomes.

As a third example, consider the experiment of throwing a die twice. A sample space for this experiment is given in Figure 2.1. Now consider the event "In two throws of a die, the score on the first throw is odd and greater than the score on the second throw." A glance at the sample space shows that the event occurs if the experiment results in 31, 32, 51, 52, 53, or 54. If the experiment results in one of the remaining 30 outcomes, the event has not occurred.

Elementary though the above illustrations are, they bring out the following important point: given an unambiguous description of an event relating to the outcome of an experiment, we can unambiguously define a subset of the sample space Ω consisting of the sample points that imply the occurrence of the event. The subset so defined has the following relation to the event: the event occurs on a given performance of the experiment if and only if the outcome of the experiment is represented by a sample point in this subset.

In developing a mathematical theory, we use the concept of a subset of the sample space to describe the intuitive notion of event. Indeed, we go a step further. Given a sample space Ω, we define the word *event* to mean a subset of Ω. So far as the mathematical theory is concerned then, an event is a subset of the sample space Ω. Bear in mind, of course, how this definition was devised through the examples above and the interpretation of the previous paragraph.

We may summarize our current position as follows. *A sample space is a set. An event is a subset of a sample space. The elements of a sample space are called sample points and are thought of as possible outcomes of a conceptual experiment. An event is a collection of sample points and is thought of as a subset of the set of all possible outcomes of a conceptual experiment.*

At this juncture, we could spend some time translating a major portion of Chapter One into the present terminology. Thus, the universal set of Chapter One is here called the sample space and subsets of the universal set are here called events. We will in fact carry across and make full use of the notation and concepts of Chapter One. For example, events will be designated by A, B, C, etc.; complements, unions, and intersections will be written as in Chapter One; and so on. In Section 2.5 we will spend some time discussing the relationships between and the operations on events, especially as they pertain to the underlying conceptual experiment.

A major benefit of translating set theory into the present framework is the ease with which we can then describe conceptual experiments. Such notation can be especially useful when we want to specify an event A

of the sample space Ω. To illustrate, let us return to the experiment of throwing a die twice, and let us take Ω to be the sample space in Figure 2.1. Consider "The score on the first throw plus the score on the second throw is an even number." The event A of Ω that corresponds to this description is $A = \{11, 13, 15, 22, 24, 26, 31, 33, 35, 42, 44, 46, 51, 53, 55, 62, 64, 66\}$. An alternative and more economical specification of A, which makes use of the nature of the sample points, is $A = \{\omega|$ the sum of the digits in ω is even$\}$. With the underlying experiment in mind, we will even give the intuitive description "The sum of the two scores is even" of this event A when no confusion can arise.

These shorthand descriptions of events of a sample space are all the more crucial when the number of sample points is large. Recall the number-of-offspring example of Section 2.3 in which the sample space had 496 sample points of the form (n_1, n_2), where n_1 and n_2 are nonnegative integers with $n_1 + n_2 \leq 30$. To specify the appropriate subset A of Ω for "More male than female offspring" would require the listing of 240 sample points. More simply, A could be written as $A = \{\omega|$ the first entry of ω exceeds the second$\}$, inasmuch as the sample point (n_1, n_2) refers to the outcome of n_1 male and n_2 female offspring. Again, when no confusion can arise, we may speak of A simply as the event of having more male than female offspring.

EXERCISES 2.4

1. Let Ω be the sample space exhibited in Figure 2.3 for the experiment of tossing a coin four times. List each of the following events of Ω.
- **(a)** Heads and tails alternate.
- **(b)** The number of heads exceeds the number of tails by one.
- **(c)** The first and third tosses have the same results.
- **(d)** The first toss results in heads.
- **(e)** The number of tails exceeds the number of heads by an even number.

Explain the answers to (b) and (e).

2. Let Ω be the sample space exhibited in Figure 2.1 for the experiment of throwing a die twice. List each of the following events of Ω.
- **(a)** The total score exceeds 6.
- **(b)** The total score is even.
- **(c)** Each throw results in an even score.
- **(d)** Each throw results in an even score larger than 2.
- **(e)** The scores on the two throws differ by 3 or more.
- **(f)** At least one throw results in an odd score.
- **(g)** At least one throw results in a score which is either odd or is 2.
- **(h)** Each throw results in a score which is either odd or is 2.
- **(i)** The sum of the squares of the scores on the two throws exceeds twice the product of two throws by 30.

What relationships can you discern between these events as subsets of Ω?

3. For the experiment given in Exercise 11 of Section 2.3, let the sample space Ω be the Cartesian product of the set $\{1, 2, 3, 4, 5, 6, 7, 8, 9, 10\}$ with itself, where the sample point (n_1, n_2) represents the outcome of a first draw of a chip numbered n_1 and a second draw of a chip numbered n_2. List each of the following events of Ω.

- **(a)** Both chips are marked with the same number.
- **(b)** The sum of the two numbers drawn is at most 7.
- **(c)** The number on the first chip exceeds 5, and the sum of the two numbers drawn is less than 11.
- **(d)** The number on the first chip exceeds 6.
- **(e)** The number on the second chip is at least as large as the number on the first.

What relations can you discern between these events as subsets of Ω? Consider the same questions for the experiment of Exercise 12 of Section 2.3 with a suitably modified sample space.

4. For the experiment given in Exercise 13 of Section 2.3, let the sample space Ω consist of all ordered pairs of letters from A, B, C, D, E, F, G, H, I, J such that no ordered pair has two entries the same, with obvious interpretation of sample points. List each of the following events of Ω.

- **(a)** The winner's name contains only one vowel.
- **(b)** The runner-up has a name in which vowels and consonants alternate.
- **(c)** The total number of letters in the names of the winner and the runner-up is less than or equal to 8.
- **(d)** The total number of letters in the names of the winner and the runner-up is less than or equal to 8, and the letter n does not occur in these names more than twice in all.
- **(e)** At least one of the winner and the runner-up has a name with a double letter in it or failing this, both have at least three vowels in their names.

What relations can you discern between these events as subsets of Ω?

5. Continue with the setup of Exercise 4. Describe the following events.

- **(a)** {(E, A), (E, G), (E, I), (E, J), (A, E), (G, E), (I, E), (J, E)}.
- **(b)** {(A, C), (B, C), (D, C), (E, C), (F, C), (G, C), (H, C), (I, C), (J, C)}.
- **(c)** {(A, C), (B, C), (D, C), (E, C), (F, C), (G, C), (H, C), (I, C), (C, J)}.

6. A marksman fires at 25 targets. Set up an appropriate sample space for this experiment, and list each of the following events.

- **(a)** He hits more targets than he misses.
- **(b)** He hits exactly 10 targets.
- **(c)** He hits 10, 11, or 12 targets.
- **(d)** He hits less than 10 or more than 12 targets.
- **(e)** He hits no targets.

7. A club has 20 junior members and 12 senior members. A president, a vice-president, a secretary, and a treasurer are to be selected. Set up an appropriate sample space, and then list each of the following events.

- **(a)** The president and the vice-president are senior members, and the secretary and the treasurer are junior members.

(**b**) The president and the vice-president are senior members.
(**c**) Either the president or the vice-president (but not both) is a junior member, and no more than two officers are junior members.
(**d**) Exactly two officers are junior members.
What modifications would you make if you were required to list the event that the oldest senior member is president?

2.5 Relations between events

Let us suppose that Ω is a fixed sample space. The events (subsets) of Ω can be inspected in the light of the set-theoretic operations and relations of Chapter One. Now, if events are viewed simply as subsets of the universal set Ω, a translation of this material into the terminology of sample space and events would surely be a thankless task. However, when a conceptual experiment underlies Ω and when the notion of the occurrence of an event is present, such a translation is rewarding. If we imagine a performance of an experiment, the phrase "The event A occurs" will mean that the experiment results in an outcome that is represented by a sample point in the subset A of Ω. Similarly, "The event A does not occur" will mean that the experiment results in an outcome that is represented by a sample point *not* in the subset A of Ω.

First, let us consider a subset A of Ω and its complementary subset A'. As we know, a sample point of Ω is in A if and only if it is *not* in A'. Phrased in terms of occurrence, complementary events stand in the following relationship: A occurs if and only if A' does not occur. A distinguished pair of complementary events is Ω and $\varnothing$. Since every sample point is in Ω and no sample points are in $\varnothing$, the event Ω always occurs while the event $\varnothing$ never occurs. Because of this, Ω is sometimes referred to as the *sure* or *certain* event while $\varnothing$ may be referred to as the *impossible* event.

If A and B are two subsets of Ω with $A \subset B$, then every sample point that is in A is also in B. In other words, if $A \subset B$, the occurrence of the event A implies the occurrence of the event B.

Given two subsets A and B of Ω, we may form the subsets $A \cup B$ and $A \cap B$ of Ω. A sample point is in $A \cup B$ if and only if it is in A, in B, or in both A and B. It follows that $A \cup B$ occurs if and only if A occurs, B occurs, or both A and B occur. The same thing can be said another way: the occurrence of the event $A \cup B$ is equivalent to the occurrence of *at least one* of the events A and B. In a similar fashion, the occurrence of the event $A \cap B$ is equivalent to the occurrence of *both* of the events A and B. Given k subsets of Ω, say $A_1, A_2, \ldots, A_k$, we may form the subsets $\bigcup_{i=1}^{k} A_i$ and $\bigcap_{i=1}^{k} A_i$ of Ω. Clearly, the occurrence of the event $\bigcup_{i=1}^{k} A_i$ is equivalent to the occurrence of *at least one* of the events $A_1, A_2, \ldots, A_k$, while the occurrence of the event $\bigcap_{i=1}^{k} A_i$ is equivalent to the simultaneous occurrence of *all* the events $A_1, A_2, \ldots, A_k$.

If A and B are disjoint subsets of Ω, then $A \cap B = \varnothing$, which may be read directly: the simultaneous occurrence of the events A and B is not possible. If $A_1, A_2, \ldots, A_k$ are mutually exclusive subsets of Ω, then $A_i \cap A_j = \varnothing$ for $i \neq j$. Here, the simultaneous occurrence of more than one of the events $A_1, A_2, \ldots, A_k$ is not possible, or, at most one of the events $A_1, A_2, \ldots, A_k$ can occur.

If A and B are subsets of Ω with $A \cup B = \Omega$, then every sample point is in A, in B, or in both A and B. This may be read as follows: the occurrence of at least one of the events A and B is certain. Similarly, if $A_1, A_2, \ldots, A_k$ are exhaustive subsets of Ω, then $\bigcup_{i=1}^{k} A_i = \Omega$. Here, the occurrence of at least one of the events $A_1, A_2, \ldots, A_k$ is certain.

To illustrate, we append a few examples.

Example 2.1

Consider the sample space for the experiment of tossing a coin four times, as displayed in Figure 2.3. We let A_1 be the event that the number of heads obtained is odd, A_2 the event that the number of tails obtained is exactly three, A_3 the event that there is at least one head obtained, A_4 the event that the second and third tosses result in tails, and A_5 the event that the number of tails obtained is even. Referring to Figure 2.3, we find $A_1 = \{\omega_2, \omega_3, \omega_4, \omega_8, \omega_9, \omega_{13}, \omega_{14}, \omega_{15}\}$, $A_2 = \{\omega_8, \omega_{13}, \omega_{14}, \omega_{15}\}$, $A_3 = \{\omega_1, \omega_2, \omega_3, \omega_4, \omega_5, \omega_6, \omega_7, \omega_8, \omega_9, \omega_{10}, \omega_{11}, \omega_{12}, \omega_{13}, \omega_{14}, \omega_{15}\}$, $A_4 = \{\omega_5, \omega_8, \omega_{13}, \omega_{16}\}$, and $A_5 = \{\omega_1, \omega_5, \omega_6, \omega_7, \omega_{10}, \omega_{11}, \omega_{12}, \omega_{16}\}$. Now, if A_2 occurs (i.e., if there are exactly three tails), then it is clear that exactly one head must appear, so that A_3 has also occurred. Thus, the occurrence of A_2 implies that of A_3, which is reflected by the relation $A_2 \subset A_3$. Similarly, $A_1 \subset A_3$. Next, we consider A_1 and A_5. If A_1 occurs (i.e., if the number of heads is odd), then the number of tails must also be odd (Why?), and then A_5 does not occur. Conversely, if A_5 occurs, A_1 does not. This is reflected by the relation $A_1 \cap A_5 = \varnothing$, which you can verify. (Indeed, you may have already observed that $A_5 = A_1'$, so that in this case $A_1 \cup A_5 = \Omega$; the events A_1 and A_5 are mutually exclusive and exhaustive.) Next, we note that $A_3 \cup A_4 = \Omega$, which reflects the fact that the occurrence of at least one of A_3 and A_4 is certain. We are saying that it is certain that *either* there will be at least one head so that A_3 occurs, *or*, that A_3 will not occur, in which case all four tosses will have resulted in tails and A_4 then occurs. Note that here A_3 and A_4 are exhaustive but all *not* mutually exclusive.

In this example, $A_1 \cap A_4 = \{\omega_8, \omega_{13}\}$. This relation is the event that both A_1 and A_4 occur, i.e., that there is an odd number of heads *and* the second and third tosses result in tails. You may wish to describe other events that can be concocted from this experiment, e.g., $A_1' \cap A_2$, $(A_1 \cup A_3)'$, etc.

Example 2.2

In connection with the granting of annuities, insurance companies are interested in the distribution of ages at death of husband and wife. The conceptual experiment of observing a couple until both husband and wife die will yield a pair of numbers (x, y), where x stands for the age of the husband at death, y for the age of the wife at death. If we agree to ignore the possibility of survival beyond age 125 and of having a married individual younger than 10 say, then a sample space Ω could consist of all ordered pairs (x, y) where x and y are integers between 10 and 125 inclusive.

Let A_1 be the event that the husband's age at death exceeds that of the wife's, and let A_2 be the event that the wife lives to at least age 65. Equivalently, $A_1 = \{(x, y) | x > y\}$ and $A_2 = \{(x, y) | y \geq 65\}$. Now $A_1 \cap A_2$ is the event $A_1 \cap A_2 = \{(x, y) | x > y \geq 65\}$ or, the event that both husband and wife live to at least age 65 and the husband's age at death is greater than the wife's age at death. Here $A_1 \cup A_2 = \{(x, y) | x > y$ or $y \geq 65$ or both $x > y$ and $y \geq 65\}$. $A_1 \cup A_2$ is the event that the husband's age at death is greater than the wife's age at death or, if not that, that the wife lives at least to age 65. $A_1' \cap A_2$ is the event that the wife's age at death is at least 65 and the husband's age at death is at most the wife's age at death.

Example 2.3

Consider the experiment of dealing a hand of 13 playing cards from a standard deck of 52 cards. These cards are divided into four suits of 13 cards each called spades, hearts, diamonds, and clubs. There are 13 denominations in each suit, 2, 3, . . . , 10, jack, queen, king, ace. Spades and clubs are black suits, while hearts and diamonds are red suits. Finally, all tens, jacks, queens, kings, and aces are called honor cards.

The number of hands that can arise when 13 cards are dealt from such a deck is rather large. Instead of attempting a formal description of all possible hands, we might imagine a number of chips, each marked with a single possible hand. This can be done in principle. Let us then take Ω to be the set of all these chips. Consider the events A_1: the hand contains at least two cards that are of denomination 10 or above, A_2: the hand contains exactly four clubs, A_3: the hand contains at least 11 hearts or diamonds, A_4: the hand contains more than two black cards, and A_5: the hand contains at least four black cards and no cards above the denomination of 10. Then, A_1 is the set of all chips marked with at least two cards of denomination 10 or above, A_2 is the set of all chips marked with exactly four clubs, and so on. In this example, we may discern the following relations: $A_5 \subset A_1'$, $A_2 \cap A_3 = \varnothing$, $A_2 \subset A_4$, $A_3 \cap A_4 = \varnothing$, and $A_3 \cup A_4 = \Omega$. No doubt you will discover others. These relations are, respectively, formalizations of the following facts, which are obvious from the descriptions given: the occurrence of A_5 implies the nonoccurrence of A_1; A_2 and A_3 cannot occur simultaneously; the occurrence of A_2 implies that of A_4; A_3 and A_4 cannot occur simultaneously; and at least one of the events A_3 and A_4 must occur.

EXERCISES 2.5

1. Let Ω be the sample space of an experiment, and let A, B, C, D, E, and F be events of Ω. Describe the following events in terms of the occurrence or nonoccurrence of some or all of the events A, B, C, D, E, and F.

(a) $A' \cap B'$.
(b) $A' \cap B$.
(c) $(A \cap B)'$.
(d) $A \cup B \cup C \cup D \cup E \cup F$.
(e) $(A' \cap B') \cup D$.
(f) $(A' \cap B) \cup (D \cap E)$.
(g) $(A \cap B') \cup (A' \cap B)$.
(h) $(A \cap B' \cap C') \cup (A' \cap B \cap C') \cup (A' \cap B' \cap C)$.
(i) $A \cap B \cap C \cap D \cap E \cap F$.
(j) $(A \cap B \cap C \cap D \cap E \cap F)'$.

2. Let Ω be the sample space of an experiment, and let A, B, and C be events of Ω. Express the following events in terms of A, B, and C using unions, intersections, and complements.

(a) A occurs, but neither B nor C occurs.
(b) At most two of the events A, B, and C occur.
(c) All three events occur.
(d) At least one of the events A, B, and C occurs.
(e) None of the three events occurs.
(f) Exactly two of the events A, B, and C occur.
(g) Exactly one of the events A, B, and C occurs, and C does not occur.
(h) A does not occur, but B does.
(i) B and C do not occur.
(j) A occurs or if not, then neither B nor C occur.

3. Let Ω be the sample space given in Figure 2.3 for four tosses of a coin. Let A be the event that exactly two heads appear, B be the event that at least two heads appear, C be the event that a head does not appear until at least one tail appears, and D be the event that the first toss results in a head.
(**a**) List the events A, B, C, and D of Ω.
(**b**) Explain any relations you see between A and B, between C and D.
Describe the events given by
(**c**) $A' \cap B'$.
(**d**) $A \cup (C \cap D)$.
(**e**) $A \cap D'$.
Write the following events in terms of A, B, C, D, and Ω.
(**f**) The first toss and at most one other toss result in tails.
(**g**) The number of heads minus the number of tails is even.
(**h**) The first toss and at most one other toss result in tails or, failing this, at least three tosses result in heads.

4. Let Ω be a sample space for n tosses of a coin. For each integer r for which $0 \leq r \leq n$, let A_r be the event that exactly r tosses result in heads. Express the following events in terms of the events $A_0, A_1, \ldots, A_n$.
(**a**) At least five tosses result in heads (assume $n \geq 5$).
(**b**) At least k tosses result in heads ($k \leq n$).
(**c**) The number of heads that appear is not less than k nor more than l ($0 \leq k \leq l \leq n$).
(**d**) The number of heads that appear is even.
(**e**) More heads than tails appear.
(**f**) The number of tails that appear is k ($k \leq n$).
Describe the events given by
(**g**) A_0'.
(**h**) $\bigcup_{i=0}^{n} A_i$.
(**i**) $A_i \cap A_j$ for $i \neq j$, $0 \leq i \leq n$, $0 \leq j \leq n$.
(**j**) $\bigcup_{i=0}^{k} A_i$ where k is the largest integer that is less than or equal to $n/2$.

5. Refer to Example 2.3, and let Ω be the sample space given there for the deal of a 13-card hand from a standard deck. Let A_1 be the event that the hand contains exactly six spades, A_2 be the event that the hand contains at least six honor cards, A_3 be the event that the number of honor cards in the hand which are not spades is at most two, A_4 be the event that the hand contains at most six cards that are not honors, and A_5 be the event that the hand contains at least one honor card that is not a spade. Write the following events in terms of A_1, A_2, A_3, A_4, and A_5 using unions, intersections, and complements.
(**a**) The hand contains exactly six honor cards.
(**b**) The hand contains at least six honor cards or, if not that, any honor cards contained in the hand are spades.
(**c**) The hand contains either one or two honor cards that are not spades.

Describe the following events.

(e) $(A_1 \cap A_2)'$.
(f) $(A_1 \cup A_4)'$.
(g) $A_4 \cap A_5$.
(h) $A_1' \cap A_4$.
(i) $(A_1 \cup A_2) \cap A_3$.

6. Let Ω be a sample space, and let A, B, and C be events of Ω. Using the language of events and occurrence, explain the following identities.

(a) $A \cup B = (A' \cap B) \cup (B' \cap A) \cup (A \cap B)$.
(b) $A = (A \cap B) \cup (A \cap B')$.
(c) $(A' \cap B) \cap (B' \cap A) = \varnothing$.
(d) $(A \cap B) \cup A' = B$.
(e) $A \cup B \cup C = (A \cap B' \cap C') \cup (A' \cap B \cap C') \cup (A' \cap B' \cap C) \cup (A' \cap B \cap C) \cup (A \cap B' \cap C) \cup (A \cap B \cap C') \cup (A \cap B \cap C)$.

CHAPTER THREE

Probability Measures

3.1 Introduction

Given an experiment and an event that may occur as a result thereof, we would like to be able to speak of the *probability* of that event. For example, we contemplate making statements like "The probability of getting three heads in five tosses of a coin is $\frac{5}{16}$," "The probability that a husband's age at death is less than his wife's age at death is $\frac{3}{5}$," "The probability of getting no aces in a bridge hand is .93," and so on. We are not concerned with numerical correctness at the moment. Rather, we wish to point out that in each case the statement refers to an event that depends on the outcome of some experiment, and that a number is being called the probability of that event. For a given experiment, many events may occur. We are going to consider an assignment of a number (probability) to each of these events. Of course, we bear in mind that for actual applications such an assignment should have some relation to reality.

In the previous chapter, we developed the mathematical framework of a sample space and its events to account for the possible outcomes of an experiment and the events that might occur. We will make the assignment of probabilities in this context. Thus, given a sample space Ω, we attach to each event A of Ω a number $P(A)$ to be called the probability of A. In mathematical terms, we speak of such an assignment as a function P that is defined on the collection of events of Ω and that takes numerical values. The function P is called a *probability measure.*

In point of fact, we will not allow arbitrary assignments of probabilities to events. We ask that certain "axioms" be satisfied. These axioms—which are set out in Section 3.2—represent formalizations of properties that are commonly held to be satisfied by "probabilities."

Further investigation will allow us to exhibit all possible probability measures on a fixed sample space. How a particular one may be selected from among these and how it is to be interpreted with respect to the underlying experiment is the subject matter of Section 3.3. In Section 3.4, the equally likely probability measure is singled out for special attention. The problems that arise there have the following simple form: if A is a given set, how many members does A have? Some useful results for this type of problem are found in the final section of this chapter.

3.2 Axioms and first consequences

Definition 3.1 A *probability measure* P on the events of a finite sample space Ω is a numerical-valued function defined on the events of Ω for which

(1) $P(A) \geq 0$ for every event A of Ω.
(2) $P(\Omega) = 1$.
(3) $P(A \cup B) = P(A) + P(B)$ for every pair of disjoint events A and B of Ω.

When A is an event from Ω, the value of a probability measure P at A, $P(A)$, is called the *probability of* A. It is interpreted as the probability of the occurrence of A in the performance of the experiment underlying Ω.

According to the first condition of Definition 3.1, probabilities of events are nonnegative. As we shall see, the second condition places all probabilities between 0 and 1, inclusive. And the consequences of the third condition will prove to be the most powerful.

Some further properties of probability measures that follow from (1), (2), and (3) are given in the following propositions. *Throughout these propositions we will suppose that* Ω *is a fixed sample space and that* P *is a fixed probability measure on the events of* Ω.

Proposition 3.1 If A is any event from Ω, then

$$P(A) + P(A') = 1$$

PROOF: A and A' are disjoint events with union Ω. Hence, by (2) and (3),

$$1 = P(\Omega) = P(A \cup A') = P(A) + P(A').$$

Since $\varnothing$ is the complement of Ω and $P(\Omega) = 1$, it must be that $P(\varnothing) = 0$.

Proposition 3.2 If A and B are two events from Ω such that $A \subset B$, then $P(A) \leq P(B)$.

PROOF: The event B may be written as $B = A \cup (B \cap A')$ when $A \subset B$. Furthermore, A and $B \cap A'$ are disjoint events. Using (3), we have $P(B) = P(A) + P(B \cap A')$. However, (1) implies $P(B \cap A') \geq 0$. So $P(A) \leq P(B)$.

Proposition 3.2, together with (1) and (2), now gives us $0 \leq P(A) \leq 1$ for any event A since we have $A \subset \Omega$ and $P(\Omega) = 1$.

Of fundamental importance is the next proposition, which enables us to write the probability of the union of any number of mutually exclusive events as the sum of the probabilities of the individual events. This seemingly stronger form of (3) is actually a direct consequence of it.

Proposition 3.3 If $A_1, A_2, \ldots, A_k$ are mutually exclusive events from Ω, then

$$P\left(\bigcup_{i=1}^{k} A_i\right) = \sum_{i=1}^{k} P(A_i)$$

PROOF: This result follows by induction. It is true for $k = 2$ by (3). If it is true for k mutually exclusive events $A_1, A_2, \ldots, A_k$,

$$P\left(\bigcup_{i=1}^{k} A_i\right) = \sum_{i=1}^{k} P(A_i)$$

Now suppose $A_1, A_2, \ldots, A_{k+1}$ are mutually exclusive. This implies that $\bigcup_{i=1}^{k} A_i$ and A_{k+1} are disjoint, and by (3)

$$P\left(\bigcup_{i=1}^{k+1} A_i\right) = P\left(\bigcup_{i=1}^{k} A_i\right) + P(A_{k+1})$$

By the induction hypothesis

$$P\left(\bigcup_{i=1}^{k} A_i\right) = \sum_{i=1}^{k} P(A_i)$$

so

$$P\left(\bigcup_{i=1}^{k+1} A_i\right) = \sum_{i=1}^{k} P(A_i) + P(A_{k+1}) = \sum_{i=1}^{k+1} P(A_i)$$

Proposition 3.3 allows us to "evaluate" the probabilities of certain events by writing these events into the form of a union of mutually exclusive events. Typical of this is the "evaluation" of $P(A \cup B)$ for two events A and B in terms of other probabilities.

Proposition 3.4 If A and B are two events from Ω, then

$$P(A \cup B) = P(A) + P(B) - P(A \cap B)$$

PROOF: We may write each of the events A, B, and $A \cup B$ as a union of mutually exclusive events from Ω, viz., $A = (A \cap B) \cup (A \cap B')$, $B = (A \cap B) \cup (A' \cap B)$, and $A \cup B = (A \cap B) \cup (A \cap B') \cup (A' \cap B)$. Applying Proposition 3.3 to these unions, we find that

$$P(A) = P(A \cap B) + P(A \cap B')$$

$$P(B) = P(A \cap B) + P(A' \cap B)$$

$$P(A \cup B) = P(A \cap B) + P(A \cap B') + P(A' \cap B)$$

Adding the expressions for $P(A)$ and $P(B)$ and comparing the sum with the expression for $P(A \cup B)$ completes the proof.

The result in Proposition 3.4 should be contrasted with (3), wherein we find $P(A \cup B) = P(A) + P(B)$ for disjoint events A and B. Proposition 3.4 expresses $P(A \cup B)$ when A and B are not necessarily disjoint. The following inequality is a consequence of Proposition 3.4 that follows by observing that $P(A \cap B)$ is nonnegative: $P(A \cup B) \leq P(A) + P(B)$ for any events A and B.

Let us turn now to the problem of specifying a probability measure P on the events of Ω. The simple-minded approach is to exhibit the entire correspondence of events to probabilities of those events. Thus, if the sample space were $\Omega = \{\omega_1, \omega_2, \ldots, \omega_n\}$, this correspondence might look like that of Figure 3.1. However, this method is inefficient, and, if the number of sample points is even moderately large, it is impractical.

Suppose, for example, that an event A is the union of three mutually exclusive events A_1, A_2, A_3. Then, according to Proposition 3.4, we have

Event	*Probability*
$\varnothing$	$P(\varnothing) = 0$
$\{\omega_1\}$	$P(\{\omega_1\})$
$\{\omega_2\}$	$P(\{\omega_2\})$
.	.
.	.
.	.
$\{\omega_n\}$	$P(\{\omega_n\})$
$\{\omega_1, \omega_2\}$	$P(\{\omega_1, \omega_2\})$
.	.
.	.
.	.
Ω	$P(\Omega) = 1$

FIGURE 3.1

$P(A) = P(A_1) + P(A_2) + P(A_3)$. Thus, if $P(A_1)$, $P(A_2)$, $P(A_3)$ are known, then $P(A)$ is completely determined. In general, if A can be somehow expressed as a union of mutually exclusive events $A_1, A_2, \ldots,$ A_k, then, once the probabilities $P(A_1), P(A_2), \ldots, P(A_k)$ are specified, they automatically determine $P(A)$.

With these remarks in mind, let us suppose that somehow we could obtain a family of events, say $A_1, A_2, \ldots, A_k$, such that they are mutually exclusive, and such that, given *any* event A in Ω, we could express A as a union of some of the events $A_1, \ldots, A_k$. Then the probability $P(A)$ is completely determined once the probabilities $P(A_1), \ldots, P(A_k)$ are specified. Indeed, if A is the union of $A_{i_1}, A_{i_2}, \ldots, A_{i_r}$, then we must have $P(A) = \sum_{j=1}^{r} P(A_{i_j})$. Of course, this means that the probability measure P can be specified completely by specifying its values on the events $A_1, \ldots, A_k$, thus effecting a considerable economy.

It is therefore natural to ask if we can find a family of events having the properties described above. The answer follows more or less immediately. Indeed, consider the simple events $\{\omega_1\}, \{\omega_2\}, \ldots, \{\omega_n\}$. These are obviously mutually exclusive. Further, if A is any event in Ω and if $A = \{\omega_{i_1}, \omega_{i_2}, \ldots, \omega_{i_r}\}$, then clearly A is the union of the simple events $\{\omega_{i_1}\}, \{\omega_{i_2}\}, \ldots, \{\omega_{i_r}\}$. The discussion above shows that if the probabilities $P(\{\omega_1\}), P(\{\omega_2\}), \ldots, P(\{\omega_n\})$ are known, then $P(A)$ can be determined for any $A \subset \Omega$.

To summarize, we have deduced that in order to specify a probability measure P on the events of Ω, it is sufficient to specify the probabilities of the simple events, say $P(\{\omega_i\}) = p_i$, $i = 1, 2, \ldots, n$. We cannot, however, be careless in specifying these numbers p_i: they must satisfy two natural restrictions. In the first place, since the numbers p_i are probabilities, we must have each p_i nonnegative. Secondly, since Ω is the union of all the simple events $\{\omega_1\}, \{\omega_2\}, \ldots, \{\omega_n\}$, the sum of all the numbers $P(\{\omega_1\}), \ldots, P(\{\omega_n\})$ must equal $P(\Omega)$ which is equal to one. Thus, we must have

$$\sum_{i=1}^{n} p_i = 1$$

If $p_1, p_2, \ldots, p_n$ are any nonnegative numbers whose sum is one, they may be used to specify a probability measure P on Ω as follows: if A is an event Ω and if $A = \{\omega_{i_1}, \omega_{i_2}, \ldots, \omega_{i_r}\}$, then we can define

$$\begin{aligned} P(A) &= \sum_{j=1}^{r} P(\{\omega_{i_j}\}) \\ &= \sum_{j=1}^{r} p_{i_j} \end{aligned}$$

Of course, $P(\varnothing) = 0$. It is easily checked that when P is thus defined, it satisfies the conditions postulated in the definition of a probability measure. The equation just written says, in words, that the probability of the event A is the sum of the probabilities of the simple events that comprise A. Furthermore, our discussion shows that every probability measure on Ω has this form.

Describing a probability measure P by means of the probabilities of the simple events is very economical. For example, suppose Ω consists of ten points $\omega_1, \omega_2, \ldots, \omega_{10}$. Then there are $2^{10} = 1024$ events of Ω and a tabulation like that of Figure 3.1 is very cumbersome. However, there are only ten simple events in Ω, and in order to specify P, all we need to do is to specify the values $p_1, \ldots, p_{10}$ of P on the ten simple events $\{\omega_1\}, \ldots, \{\omega_{10}\}$, respectively. This is clearly a considerable economy. Of course, $p_1, \ldots, p_{10}$ must be nonnegative and must have a sum of one.

It will be seen shortly that a variety of problems may be tackled with this technique of determination of P.

EXERCISES 3.2

1. Which of the following are probability measures on the events of $\Omega = \{\omega_1, \omega_2, \omega_3\}$?

Event	P_1	P_2	P_3	P_4	P_5	P_6
$\varnothing$	0	0	0	0	0	0
$\{\omega_1\}$	$\frac{1}{4}$	$\frac{1}{4}$	$\frac{1}{3}$	$\frac{1}{2}$	$\frac{1}{2}$	$\frac{1}{2}$
$\{\omega_2\}$	$\frac{1}{4}$	$\frac{1}{2}$	$\frac{1}{3}$	$\frac{1}{4}$	0	$\frac{1}{4}$
$\{\omega_3\}$	$\frac{1}{2}$	$\frac{1}{4}$	$\frac{1}{3}$	$\frac{1}{8}$	$\frac{1}{2}$	$\frac{1}{4}$
$\{\omega_1, \omega_2\}$	$\frac{3}{8}$	$\frac{3}{4}$	$\frac{2}{3}$	$\frac{3}{4}$	$\frac{1}{2}$	$\frac{5}{4}$
$\{\omega_1, \omega_3\}$	$\frac{1}{2}$	$\frac{1}{2}$	$\frac{2}{3}$	$\frac{5}{8}$	1	$\frac{3}{4}$
$\{\omega_2, \omega_3\}$	$\frac{3}{4}$	$\frac{3}{4}$	$\frac{2}{3}$	$\frac{3}{8}$	$\frac{1}{2}$	$\frac{1}{2}$
Ω	1	1	1	1	1	1

2. Determine the probability measure P at each of the events of $\Omega = \{\omega_1, \omega_2, \omega_3, \omega_4\}$, if

(a) $P(\{\omega_1\}) = \frac{1}{8}$, $P(\{\omega_2\}) = \frac{1}{8}$, $P(\{\omega_3\}) = \frac{3}{8}$, $P(\{\omega_4\}) = \frac{3}{8}$.
(b) $P(\{\omega_1\}) = \frac{2}{5}$, $P(\{\omega_2\}) = \frac{3}{10}$, $P(\{\omega_3\}) = \frac{1}{5}$.
(c) $P(\{\omega_1\}) = \frac{1}{6}$, $P(\{\omega_2\}) = \frac{1}{2}$, $P\{(\omega_3\}) = 0$, $P(\{\omega_4\}) = \frac{1}{3}$.

3. Suppose that P is a probability measure on the events of Ω. Find $P(A')$, $P(A' \cap B)$, $P(A \cap B')$, $P(A \cup B)$, and $P(A' \cap B')$ when

(a) $P(A) = \frac{1}{3}$, $P(B) = \frac{1}{4}$, $P(A \cap B) = \frac{1}{6}$.
(b) $P(A) = \frac{1}{2}$, $P(B) = \frac{1}{8}$, $P(A \cap B) = \frac{1}{8}$.
(c) $P(A) = \frac{1}{2}$, $P(B) = \frac{1}{8}$, $P(A \cap B) = 0$.
(d) $P(A) = 1$, $P(B) = \frac{1}{4}$, $P(A \cap B) = \frac{1}{4}$.

4. Suppose P is a probability measure on the events of Ω with $P(A) = \frac{3}{4}$ and $P(B) = \frac{5}{8}$. Show that
   (a) $P(A \cup B) \geq \frac{3}{4}$. (b) $\frac{3}{8} \leq P(A \cap B) \leq \frac{5}{8}$.
   (c) $\frac{1}{8} \leq P(A \cap B') \leq \frac{3}{8}$.
   Can you find instances—i.e , by choice of P, Ω, A, and B—in which the various equalities like $P(A \cap B) = \frac{5}{8}$ hold?

5. Suppose P is a probability measure on the events of Ω with $P(A) = \frac{1}{2}$ and $P(B) = \frac{2}{3}$. Give inequalities analogous to those in Exercise 4 for $P(A \cup B)$, $P(A \cap B)$, and $P(A' \cup B')$.

6. Suppose P is a probability measure on the event of Ω. Show for events A, B, and C of Ω that

$$P(A \cup B \cup C) = P(A) + P(B) + P(C) - P(A \cap B)$$
$$- P(A \cap C) - P(B \cap C) + P(A \cap B \cap C)$$

7. Let $\Omega = \{\omega_1, \omega_2, \ldots, \omega_n\}$, and let $p_1, p_2, \ldots, p_n$ be nonnegative numbers with $\sum_{i=1}^{n} p_i = 1$. Show directly that if P is defined on the events of Ω by $P(\varnothing) = 0$ and $P(\{\omega_{i_1}, \omega_{i_2}, \ldots, \omega_{i_k}\}) = \sum_{j=1}^{k} p_{i_j}$ for any selection $\omega_{i_1}, \omega_{i_2}, \ldots,$ ω_{i_k} of one or more sample points, then P is a probability measure on the events of Ω.

8. Show that if P is a probability measure on the events of Ω and if $A_1, A_2, \ldots,$ A_k are events of Ω, then

$$P(\bigcup_{i=1}^{k} A_i) \leq \sum_{i=1}^{k} P(A_i)$$

9. Suppose P is a probability measure on the events of Ω. An event A of Ω is called P-*null* if $P(A) = 0$. Show that if both A and B are P-null, so are the events $A \cap B$, $A \cup B$, and $A \cap B'$. Show that if A is P-null and B is any event, then $P(A \cap B) = 0$ and $P(A \cup B) = P(B)$.

10. Suppose P is a probability measure on the events of Ω. An event A of Ω is called P-*sure* if $P(A) = 1$. Show that if both A and B are P-sure, so are the events $A \cup B$, $A \cap B$, and $A \cup B'$. Show that if A is P-sure and B is any event, then $P(A \cup B) = 1$ and $P(A \cap B) = P(B)$.

11. Suppose P is a probability measure on the events of Ω. Show that if there is a total of k P-null events of Ω, then there are also k P-sure events of Ω.

*12. Suppose P is a probability measure on the events of Ω. Show that
   (a) There is a unique P-null event D of Ω that contains every P-null event.
   (b) There is a unique P-sure event E of Ω that is contained in every P-sure event.
   (c) $D' = E$.

---

* Starred exercises either are more difficult than the standard exercises in this book or require some knowledge of the calculus.

***13.** State and prove an analogue to Proposition 3.4 and Exercise 6 for the determination of $P(\bigcup_{i=1}^{k} A_i)$, k being any positive integer.

3.3 Probability interpretation

The major portion of this book deals with deductions akin to the following: if the probabilities of some events are such and such, then the probability of some other event is so and so. At this point, we pause to discuss the usual meaning attached to the probability of an event as it relates to a real experiment. The interpretation of probability is intertwined with the problem of how and when to choose a probability measure on a sample space as a reflection of reality. We set out some fundamentally different situations. It should be mentioned initially that there is a wide spectrum of opinion as to which of these situations allow a proper application of the mathematical theory.

Let us begin by imagining an experiment that can be performed a number of times under similar conditions. Such an experiment might consist of flipping a coin, rolling a die, playing a game of blackjack, calculating gas mileage under normal driving conditions, observing the yearly sex ratios of newborn children in the United States, tabulating the crop in bushels from a given plot of land, etc. These examples typify experiments in which outcomes vary between experiments despite carefully controlled conditions. For example, when flipping a coin, we do not expect to be able to predict exactly when heads will or will not occur, even if we attempt to apply always the same amount of impetus to the coin and try also to keep the landing surface the same. We suppose, however, that the possible outcomes of the experiment in question can be agreed upon and then represented by the sample points $\omega_1, \omega_2, \ldots, \omega_n$. If the experiment is run a total of T times, we may find that ω_1 is the outcome T_1 of these times, ω_2 is the outcome T_2 of these times, $\ldots$, ω_n is the outcome T_n of these times, $T_1 + T_2 + \cdots + T_n = T$. The *relative frequency* of ω in this sequence of experiments is defined to be T_i/T, $i = 1, 2, \ldots, n$.

The basic empirical fact about such experiments is the following. Suppose we perform a sequence of T experiments and note the relative frequency of, say, the outcome ω_1. The relative frequency will be a number, say f_1, between 0 and 1, which will vary from one sequence to another; i.e., if we perform another sequence of T experiments and find the relative frequency of the outcome ω_1, we shall in general get a different value for f_1. However, it can be observed empirically, that if T is large, the variations in the relative frequency f_1 from one sequence to the next are small. That is, the values of f_1 that we obtain from different sequences of length T tend to cluster together for large T. It is but one step from this to postulate that there is some ideal value p_1 about which this clustering takes place. Of course, the same may then be said of the relative frequency f_i of the outcome ω_i, $i = 2, 3, \ldots, n$. Thus, we think of postulating the existence of ideal values p_i, $i = 1, 2, \ldots, n$, around which the relative frequencies f_i, $i = 1, 2, \ldots, n$, tend to cluster. Inasmuch as $f_1 + f_2 + \cdots + f_n = 1$, we take these values also to satisfy $p_1 + p_2 + \cdots + p_n = 1$. The standard choice of a probability measure P on the events of Ω for this type of experiment is now determined by setting $P(\{\omega_i\}) = p_i$, $i = 1, 2, \ldots, n$.

For purposes of illustration, the results in Figure 3.2 were obtained in six sequences of coin flips, each of length ten, followed by six sequences, each of length twenty-five, and then by six sequences, each of length 200. The corresponding relative frequencies of heads are given in Figure 3.3. With $T = 10$, these relative frequencies vary between .3 and .7, with $T = 25$ between .44 and .6, and with $T = 200$ between .46 and .51. An assignment of probability $\frac{1}{2}$ to the event {heads} and of probability $\frac{1}{2}$ to the event {tails} would represent an amalgamation of the experience gained in flipping this coin, since the relative frequencies of heads are collecting around $\frac{1}{2}$. We have chosen $\frac{1}{2}$ for the sake of simplicity and

Number of heads observed

Length \ Sequence	1	2	3	4	5	6
10	3	4	7	6	4	5
25	14	11	15	14	13	15
200	98	102	97	95	101	92

FIGURE 3.2

Frequency of heads observed

Length \ Sequence	1	2	3	4	5	6
10	.3	.4	.7	.6	.4	.5
25	.56	.44	.6	.56	.52	.6
200	.49	.51	.485	.475	.505	.46

FIGURE 3.3

because it fits well with what we have seen in our (limited) experience. One could as well argue that an assignment of probability .49 to {heads} is a synopsis of this experience. A choice of .43 would by now, however, no longer seem reasonable.

The frequency interpretation of the statement "The probability of the event A is p" is as follows: if the underlying experiment is run a large number T of times under similar conditions, we can expect the number of occurrences of A to be approximately $p \cdot T$.

Similar reasoning based on birth statistics of very large size (i.e., large numbers of births), lead statisticians to assign a probability of approximately .505 to {male birth} and of approximately .495 to {female birth}. For practical application, these probabilities mean different things to different people. For example, a baby clothes manufacturer might decide to make slightly more blue clothes than pink. On the other hand, an expectant father may view an upcoming birth as a special experiment that cannot be run a large number of times under such similar conditions as holding the parents constant. In the same vein, insurance companies can make excellent use of the relative frequencies of events having the form {a healthy male of age x survives to age y} over large numbers of experiments. However, a prospective policy buyer may view these relative frequencies with some skepticism for his own situation and so question whether the policy is or is not a good buy.

These considerations lead us in the direction of experiments that cannot be performed many times under similar conditions, at least from some person's standpoint. The experiment might be attempting to hit the moon with a missile, passing or failing a college course, getting through the next month without major mishap, guessing the winner in a given horse race, etc. Can we make use of our experience with respect to other experiments of a related nature in order to talk of probabilities of events? Suppose, for example, the coin of Figure 3.2 is discarded and we wish to flip another coin. Is it proper to ascribe probability $\frac{1}{2}$ to {heads} without benefit of prior experimentation? The best answer to this question is that, "It depends." Let us consider two extremes. We are aware of the workmanship put into a newly minted coin and that rarely is one produced with a striking imbalance. Given such a coin, we might well expect that if flipped a large number of times, roughly half of the flips would result in heads. On the other hand, a coin presented to us by a stranger who offers to pay us \$100 every time we can flip heads and to collect \$1 every time we flip tails would call forth other relevant experience—it doesn't pay to fool with strangers, all that glitters is not gold, etc.

For experiments in which prior information of the type that could produce long-run relative frequencies is not directly available, disagreement exists over whether an assignment of probabilities should be made.

One pertinent question is this: can useful results be obtained by making an assignment as opposed to not doing so? The answer depends on the use to be made of the results and the consequences of that use. As a simple illustration, if betting with the stranger is contemplated and probability $\frac{1}{2}$ is assigned to {heads}, we would decide betting is very profitable. If the coin has two tails, a serious mistake has been made.

EXERCISES 3.3

1. Suppose $\Omega = \{\omega_1, \omega_2, \ldots, \omega_n\}$. An experiment with Ω as a sample space is run a total of T times of which T_1 result in the outcome ω_1, T_2 result in the outcome $\omega_2, \ldots, T_n$ result in the outcome ω_n, $T_1 + T_2 + \cdots + T_n = T$. Show that if $T(A)$ denotes the number of times the event A occurs for each event A of Ω, then the function P defined by $P(A) = T(A)/T$ for every event A of Ω is a probability measure on the events of Ω. What form does this probability measure take if $T = 1$? If $T = 2$?

2. Flip a coin through six sequences of 50 tosses each. Do the observed relative frequencies of heads support a frequency interpretation of "The probability of heads is $\frac{1}{2}$"? Explain.

3. Comment on the similarity (or lack thereof) of conditions under which the following sequences of experiments are run.

(**a**) The daily receipts of a department store are tabulated to the nearest hundred of dollars over a period of 3 years.

(**b**) Traffic deaths per week in the United States are recorded over a period of 200 weeks.

(**c**) The number of stocks traded on the New York Stock Exchange is noted daily for a period of 2 years; for a period of 20 years.

(**d**) The number of stocks on the New York Stock Exchange that advance in price is noted daily for a period of 2 years; for a period of 20 years.

(**e**) The starting positions of winning horses are tabulated throughout a season of races.

(**f**) The number of misprints per page of the *New York Times* is tabulated for a period of 10 weeks; for a period of 10 years.

(**g**) The number of want ads in the *New York Times* per day is recorded for 2 years; for 20 years.

4. Roll a die through six sequences of 50 throws each. Do the observed relative frequencies support a frequency interpretation of "The probability of a score of i is $\frac{1}{6}$" for each of $i = 1, 2, 3, 4, 5, 6$?

5. Roll two dice through six sequences of 50 throws each. Do the observed relative frequencies support a frequency interpretation of "The probability of a total score of 11 is $\frac{1}{18}$"?

3.4 The equally likely probability measure

Let $\Omega = \{\omega_1, \omega_2, \ldots, \omega_n\}$. Given any nonnegative numbers $p_1, p_2, \ldots, p_n$ of which the sum is 1, we know that a probability measure P on the events of Ω would be determined by setting $P(\{\omega_i\}) = p_i$, $i = 1, 2, \ldots, n$. In particular, a measure P can be defined by choosing $p_i = 1/n$, $i = 1, 2, \ldots, n$. This measure P has the following simple description: if A is an event from Ω and $N(A)$ is the number of sample points in A, then $P(A) = N(A)/n = N(A)/N(\Omega)$. The fact that each simple event has the same probability is emphasized by calling this probability measure the *equally likely* measure. Equivalently, whenever it is convenient, the outcomes will be called equally likely.

Why single out this probability measure? There are two reasons. On the one hand, it is a natural selection for a variety of experimental situations (the roll of a balanced die, the draw of a card, etc.). Indeed, in many cases great pains are taken to insure that it is a good approximate model for the experiment (witness, for example, an expensive roulette wheel or a barrel churn for drawing lucky numbers). On the other hand, and somewhat selfishly, this probability measure is easy to describe and to handle.

In the wake of the choice of an equally likely probability measure is a multitude of problems that, when suitably abstracted, have the form: if P is the equally likely measure on a sample space Ω and if A is an event from Ω, determine $P(A)$. Since $P(A) = N(A)/N(\Omega)$, we must determine $N(A)$ [and quite possibly $N(\Omega) = n$]. To illustrate these (aptly termed) counting problems and at the same time to give a small indication of the degree of difficulty that can be reached, we turn to a number of examples. The first two examples below are straightforward counting problems and so are solved immediately. The latter three examples typify counting problems that are solved in the next section.

Example 3.1

A card is drawn from an ordinary deck of 52 cards. A suitable sample space is $\Omega = \{AS, KS, QS, \ldots, 4C, 3C, 2C\}$. Suppose P is the equally likely measure on Ω. Let us determine the probability of drawing an ace and the probability of drawing a red queen. No special difficulty arises here. In the first instance, $N(\{\omega | \omega \text{ is an ace draw}\}) = 4$, while, in the second, $N(\{\omega | \omega \text{ is a red queen draw}\}) = 2$. The probabilities are, respectively, $\frac{4}{52} = \frac{1}{13}$ and $\frac{2}{52} = \frac{1}{26}$.

Example 3.2

A number is drawn from a hat that contains the numbers 1, 2, . . . , 50. Let P be the equally likely measure on $\Omega = \{1, 2, \ldots, 50\}$. What then are the probabilities of drawing, respectively, a number divisible by 4, a number that divides 66, a prime number? A number divisible by 4 must be $4k$ for some integer k. In the present case, such numbers come from $k = 1, 2, \ldots, 12$, and so the first probability is $\frac{12}{50} = \frac{6}{25}$. We may write 66 as $2 \cdot 3 \cdot 11$, and a number that divides 66 must be a product of some of these factors or it must be 1. There are seven such draws, and the probability is $\frac{7}{50}$. For the last case, a listing gives $\{2, 3, 5, 7, 11, 13, 17, 19, 23, 29, 31, 37, 41, 43, 47\}$, and therefore the probability is $\frac{15}{50} = \frac{3}{10}$.

Example 3.3

First, second, and third prize winning tickets are drawn from a bowl full of tickets numbered 1, 2, . . . , 540. A suitable sample space consists of ordered triples of integers through 540, no two entries in any triple the same. Suppose P is the equally likely measure on this sample space. What are, respectively, the probabilities that ticket number 1 wins first prize, that ticket number 1 wins first prize and ticket number 2 wins second prize, that ticket number 1 wins some prize?

Example 3.4

A committee of six is to be drawn from a group consisting of 25 men and 15 women. Supposing no two people have the same name, a sample space could consist of all distinct six-tuples of names, no six-tuple having the same name more than once. Suppose P is equally likely on this sample space. What are, respectively, the probabilities that the tallest man will be on the committee, that the tallest man and the shortest woman will be on the committee, that the committee will have five women members?

Example 3.5

Mr. Adams, Mr. Brown, and Mr. Clay sit down to play a hand of poker. Accordingly, each is dealt a hand of five cards from an ordinary deck of 52. Suppose all possible deals are equally likely. What are, respectively, the probabilities that Mr. Adams is dealt four aces, that somebody is dealt four aces, that Mr. Adams is dealt four aces and Mr. Brown is dealt four kings?

The latter three examples, although superficially difficult, are indicative of only the most rudimentary counting problems. The essential elements of their solutions, however, can be put to use in further and more taxing problems. These matters are taken up in Section 3.5.

EXERCISES 3.4

1. Suppose P is the equally likely measure on the events of $\Omega = \{x | \, x \text{ is an integer between 1 and 100 inclusive}\}$. Find the probability of the following events of Ω.
(**a**) $\{x | \, x \text{ is divisible by } 3\}$.
(**b**) $\{x | \, x \text{ is divisible by } 13\}$.
(**c**) $\{x | \, x = 7k + 4 \text{ for some positive integer } k\}$.
(**d**) $\{x | \, x \text{ is a prime number}\}$.
(**e**) $\{x | \, 120 \text{ is divisible by } x\}$.
(**f**) $\{x | \, x^2 < 185\}$.

2. Suppose a die is thrown twice and the 36 outcomes of Figure 2.2 are taken to be equally likely. Find the probability of the following events.
(**a**) The first score is odd.
(**b**) The first score is odd, and the second score is even.
(**c**) The total score is 11.
(**d**) The first score is smaller than the second.
(**e**) The first score is larger than the second score plus 2.

3. Suppose a coin is flipped four times and the 16 outcomes of Figure 2.3 are taken to be equally likely. Find the probability of the following events.
(**a**) The first and second tosses are heads.
(**b**) The first toss is heads, and the second toss is tails.
(**c**) More heads than tails are tossed.
(**d**) All four tosses are the same.
(**e**) Any head that is tossed precedes every tail that is tossed.

4. A man notes that after a purchase his coin change is equally likely to be in the amount of 0, 1, 2, . . . , 99 cents. If change is made first with as many quarters as possible, then with as many dimes as possible, then with as many nickels as possible, then pennies, find the probability that among his change he receives the following.
(**a**) Exactly two quarters.
(**b**) Exactly two quarters and exactly two dimes.
(**c**) Exactly two dimes.
(**d**) Exactly three pennies and exactly one nickel.
(**e**) At least one quarter.
(**f**) At least two pennies.
(**g**) No dimes.

5. We will call any sequence of letters a word. How many four-letter words may be formed from the letters of the word "part"? From the letters of the word "apart"?

6. A television network must present five commercials, one commercial three times and another twice. In how many orders may these be given? In how many orders may they be given if the two of one kind are not to be given consecutively?

7. How many different committees of size two may be formed from a group of four people? How many different committees of size two may be formed from a group of four people if one committeeman is designated as chairman?

8. A comedian does routines on television guest appearances, two routines to a show. How many routines must he have for four shows if he does not wish to use the same two routines on different shows and does not wish to use any one routine on consecutive shows?

9. In how many ways may four people be seated at a table marked with positions north, south, east, and west? In how many ways may they be seated if two ways are regarded as the same if every person has the same left-hand neighbor and the same right-hand neighbor?

3.5 Counting

As the examples in Section 3.4 have indicated, it is necessary to develop certain counting techniques with which to solve some rather easily stated problems in the case of an equally likely probability measure. Our actual requirements are few in this respect, and we will not delve into really sophisticated counting here.

The Multiplication Principle The basic technique for counting, of which liberal use is made in this section, is the multiplication principle. Let us imagine a job that is to be accomplished in two stages. We suppose the first stage may be done in k different ways and then, when this stage is completed, there are m ways to do the second stage. The multiplication principle states that the number of ways in which the whole job may be done is $k \cdot m$. The same observation is easily extended by induction to jobs of r stages. Thus, a job that consists of r stages and is accomplished in one of n_1 ways at the first stage, then in one of n_2 ways at the second stage, . . . , then in one of n_r ways at the rth stage, may be done in $n_1 \cdot n_2 \cdot \cdots \cdot n_r$ ways overall.

Permutations Suppose we consider a collection of k distinguishable objects, for example, the integers 1, 2, . . . , k. We let $r \leq k$. An arrangement of r of these k items in an ordered list (i.e., in a list in which the objects are ordered as first, second, . . . , rth) is called a *permutation* of k

objects taken r at a time. The number of possible permutations in this case is denoted by $(k)_r$.

We find an expression for $(k)_r$ by viewing the construction of a permutation of k objects taken r at a time as a job of r stages. For the first stage, an object is designated to be the first object in the arrangement, and this may be done in one of k ways. After the first stage is completed, another object is designated to be the second, and this may now be done in one of $k - 1$ ways. Continuing in this way, once the $(r - 1)$st stage is completed, an object not previously chosen is designated to be the rth, and this may be done in one of $k - r + 1$ ways. According to the multiplication principle, we have

$$(k)_r = k(k-1)\cdots(k-r+1) = \frac{k!}{(k-r)!} \tag{3.1}$$

Example 3.6

We return to Example 3.3. There are $(540)_3$ sample points in the sample space. First, we are to determine the probability the ticket number 1 wins first prize. Thus, we must find the number of sample points that list ticket number 1 in the first position. There are as many of these as there are ways to order the remaining 539 tickets into two positions, viz., $(539)_2$. Therefore, since the total number of triples is $(540)_3$, the probability that ticket number 1 is the winning ticket is

$$\frac{(539)_2}{(540)_3} = \frac{539!/537!}{540!/537!} = \frac{1}{540}$$

In the same way, we find the number of sample points listing ticket number 1 in the first position and ticket number 2 in the second position to be the number of ways the remaining tickets may be ordered into the third position, viz., $(538)_1 = 538$. The probability required is then

$$\frac{538}{(540)_3} = \frac{538}{540 \cdot 539 \cdot 538} = \frac{1}{540 \cdot 539}$$

That ticket number 1 wins some prize means ticket number 1 wins first, second, or third prize. It may be seen that these are mutually exclusive events, each having probability $\frac{1}{540}$. The probability that ticket number 1 wins some prize must then be $\frac{3}{540} = \frac{1}{180}$.

Combinations Let us continue with a collection of k distinguishable objects. We let $r \leq k$. A selection of r different objects from among these k is called a *combination* of k objects taken r at a time. The number of possible combinations in this case is denoted by $\binom{k}{r}$.

To determine $\binom{k}{r}$, consider the task of forming all permutations of k objects taken r at a time. We know that this task can be done in $(k)_r$ ways. Alternativly, we can regard the task as being performed in two stages. In the first stage, we merely select r objects from the k given objects, which may be done in $\binom{k}{r}$ ways. In stage two, we arrange the r selected objects in all possible permutations, which may be done in $(r)_r = r!$ ways. By the multiplication principle, the entire task can be performed in $\binom{k}{r} \cdot r!$ ways. Thus this number must equal $(k)_r$, i.e., $\binom{k}{r} \cdot r! = (k)_r$, so

$$\binom{k}{r} = \frac{(k)_r}{r!} = \frac{k!}{(k-r)!r!} \tag{3.2}$$

Example 3.7

The number of committees of six people from 40 people as required in Example 3.4 is $\binom{40}{6}$. The number of committees on which the tallest man is a member is the number of committes of five that may be formed from 39 people, viz., $\binom{39}{5}$. The first probability in question is

$$\frac{\binom{39}{5}}{\binom{40}{6}} = \frac{39!}{34!5!} \cdot \frac{34!6!}{40!} = \frac{6}{40} = \frac{3}{20}$$

Similarly, the number of committees having both the tallest man and shortest woman as members is the number of those having four members drawn from 38. This probability is

$$\frac{\binom{38}{4}}{\binom{40}{6}} = \frac{38!}{34!4!} \cdot \frac{34!6!}{40!} = \frac{6 \cdot 5}{40 \cdot 39} = \frac{1}{52}$$

Committees having five women and one man as members may be formed in two stages: first choose the women in one of $\binom{15}{5}$ ways and then the man in one of $\binom{25}{1} = 25$ ways. The probability of having five women on the committee is then

$$\frac{25 \cdot \binom{15}{5}}{\binom{40}{6}} = \frac{25 \cdot 15!}{10!5!} \cdot \frac{34!6!}{40!} = \frac{150 \cdot 15!34!}{10!40!}$$

Permutations and Combinations Continued We next consider k objects not all of which are distinguishable. They might be colored balls of uniform size, for example, so that balls of the same color cannot be told apart while those of different color can. We suppose there are k_1 of one kind and indistinguishable, k_2 of another kind and indistinguishable, ... , k_q of yet another kind and indistinguishable, $k_1 + k_2 + \cdots + k_q = k$. Let us determine the number of distinct permutations of all k objects, say, $(k)_{k_1,k_2,\ldots,k_q}$. If the balls were distinguishable, this number would be $k!$ However, let's consider one of these $k!$ permutations. We have allowed the objects of the first kind to be ordered among the positions they occupy in $k_1!$ ways, the objects of the second kind to be ordered among the positions they occupy in $k_2!$ ways, ... , the objects of the qth kind to be ordered among the positions they occupy in $k_q!$ ways. The permutation we have singled out has been counted $k_1!k_2!\cdots k_q!$ times among all permutations. Therefore,

$$(k)_{k_1,k_2,\ldots,k_q} = \frac{k!}{k_1!k_2!\cdots k_q!} \tag{3.3}$$

To view this problem from the opposite direction, let's suppose we must place k distinguishable objects in q boxes, k_1 in the first box, k_2 in the second box, ... , k_q in the qth box, $k_1 + k_2 + \cdots + k_q = k$. To determine the number of ways in which this may be done, we imagine the permutations above as the assignment of positions (new distinguishable objects) to kinds of items (boxes). The distinguishable objects, k in number, must be assigned to q boxes, k_1 to the first, k_2 to the second, ... , k_q to the qth. According to this interpretation, the number of ways this may be done is again $k!/k_1!k_2!\cdots k_q!$. This last fraction is written more commonly as $\binom{k}{k_1,\ k_2,\ \ldots,\ k_{q-1}}$. It is also, as seen above, the number of ways of combining k distinguishable objects into groups of size k_1, k_2, ... , k_q.

Example 3.8

Now Example 3.5 can be handled. First, the number of deals of five cards to each of three players is the number of ways to form 52 distinguishable objects into groups of 5, 5, 5, and 37; thus, it is $\binom{52}{5,\ 5,\ 5}$. To find the probability that Mr. Adams has four aces, note that there are $\binom{48}{1} = 48$ hands that have four aces (four aces and one other card). For the first stage, assign one of these hands to Mr. Adams. For the second stage, combine 47 cards into groups of 5, 5, and 37 in one of $\binom{47}{5,\ 5}$ ways. The total number of appropriate hands is $48 \cdot \binom{47}{5,\ 5}$, and the required probability is

$$\frac{48 \cdot \binom{47}{5,\ 5}}{\binom{52}{5,\ 5,\ 5}} = \frac{48 \cdot 47!}{37!5!5!} \cdot \frac{37!5!5!5!}{52!} = 48 \cdot \frac{5!47!}{52!}$$

This must be the same as the probability of Mr. Brown or of Mr. Clay having four aces, and the event that someone has four aces in a union of these mutually exclusive events. Thus, the probability that someone has four aces is $3 \cdot 48 \cdot (5!47!/52!)$.

Lastly, we must construct deals in which Mr. Adams has four aces and Mr. Brown has four kings. Let us assign the aces to Mr. Adams, the kings to Mr. Brown, and combine the 44 remaining cards into groups of one (Mr. Adam's other card), one (Mr. Brown's other card), five (Mr. Clay's hand), and 37. This may be done in $\binom{44}{1,\,1,\,5}$ ways. The probability in question is

$$\frac{\binom{44}{1,\,1,\,5}}{\binom{52}{5,\,5,\,5}} = \frac{44!}{37!5!} \cdot \frac{37!5!5!5!}{52!} = \frac{5!5!44!}{52!}$$

We conclude this section with two examples of a slightly different nature.

Example 3.9

Occupancy Suppose k balls are successively placed in q boxes numbered $1, 2, \ldots, q$. The first ball may be placed in one of q ways, the second in one of q ways, ..., the kth in one of q ways. Let us suppose P is equally likely over all q^k placements. We determine the probability that

(1) Box number 1 receives k_1 balls.
(2) Box number 1 receives k_1 balls, box number 2, k_2 balls, ..., box number q, k_q balls.
(3) No box receives more than one ball.

For (1), k_1 balls may be chosen from k in one of $\binom{k}{k_1}$ ways for placement in the first box. The remaining balls may be placed in $q - 1$ boxes in one of $(q - 1)^{k-k_1}$ ways. The probability that box number 1 receives k_1 balls is

$$\binom{k}{k_1} \frac{(q-1)^{k-k_1}}{q^k} = \binom{k}{k_1} \left(\frac{1}{q}\right)^{k_1} \left(1 - \frac{1}{q}\right)^{k-k_1}$$

The k balls may be formed into groups of $k_1, k_2, \ldots, k_q$ in one of $\binom{k}{k_1,\, k_2,\, \ldots,\, k_{q-1}}$ ways and then assigned to boxes $1, 2, \ldots, q$, respectively. The probability of this arrangement is therefore

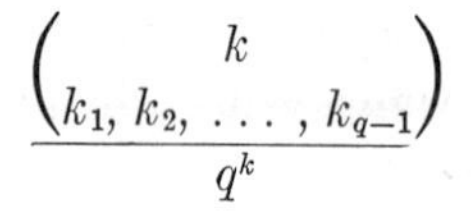

$$\frac{\binom{k}{k_1, k_2, \ldots, k_{q-1}}}{q^k}$$

In order that no box will receive more than one ball, we must have no more balls than boxes, $k \leq q$. Now for the event to occur, the first ball may be placed in one of q boxes. The second ball may be placed in one of $q - 1$ boxes, . . . , the kth ball may be placed in one of $q - k + 1$ boxes. There are $q!/(q - k)!$ ways of doing this and the probability is $\dfrac{q!}{(q - k)!} \dfrac{1}{q^k}$.

Example 3.10

Matching Suppose two decks of n cards are numbered 1 through n. Suppose, after shuffling, they are turned face up simultaneously one at a time. Let a match be counted each time the cards turned over possess the same number. We discuss the problem of finding the probability of k matches, $0 \leq k \leq n$, when all orderings are equally likely. The first reduction that can be made is to suppose that one deck gives successively $1, 2, \ldots, n$ and that the other deck is matched against this; for, if this were not the case, we could renumber all cards according to the order in which they appear in the first deck. Thus, a suitable sample space has in it $n!$ sample points and we suppose that P is equally likely on it. We consider first the probability of having no matches. When there are n cards, let us call this probability p_n, so $p_n = N_n/n!$, where N_n is the number of orderings in which no matches occur.

To derive p_n, we obtain a relation between adjoining values of p_n. To construct appropriate orderings, suppose we begin by placing a card numbered other than 1 in the first place, which can be done in $n - 1$ ways, and suppose we choose the card numbered r. To complete an ordering with no matches, there are two ways to proceed. First we might put card number 1 in the rth place, in which case we must order the cards numbered $2, 3, \ldots, r - 1, r + 1, \ldots, n$ in the same numbered places without a match, which can be done in N_{n-2} ways. Second, we might prevent the card numbered 1 from falling in the rth place while ordering the cards numbered $1, 2, \ldots, r - 1, r + 1, \ldots, n$ into the places 2, $3, \ldots, n$, which can be done in N_{n-1} ways. It then must be that $N_n = (n - 1)(N_{n-1} + N_{n-2})$. Dividing both sides of this equation by $n!$, we find that this is the same as

$$p_n = \left(\frac{n-1}{n}\right) p_{n-1} + \left(\frac{1}{n}\right) p_{n-2} \quad \text{or} \quad p_n - p_{n-1} = \left(\frac{-1}{n}\right)(p_{n-1} - p_{n-2})$$

This relationship between values of p_n says then that each successive difference of values is an appropriate fraction of the previous difference. If this is carried backwards, we find

$$p_n - p_{n-1} = \left(\frac{-1}{n}\right)\left(\frac{-1}{n-1}\right)(p_{n-2} - p_{n-3})$$
$$= \left(\frac{1}{n(n-1)}\right)\left(\frac{-1}{n-2}\right)(p_{n-3} - p_{n-4}) \text{ etc.}$$

Now $p_2 - p_1 = \frac{1}{2}$, and therefore $p_n - p_{n-1} = (-1)^n/n!$. Finally,

$$p_n = p_1 + (p_2 - p_1) + (p_3 - p_2) + \cdots + (p_n - p_{n-1})$$
$$= (\tfrac{1}{2}) - (\tfrac{1}{6}) + (\tfrac{1}{24}) - \cdots + \frac{(-1)^n}{n!}$$

since p_1 is in fact zero. Successively, we have $p_2 = \frac{1}{2}$, $p_3 = \frac{1}{3}$, $p_4 = \frac{3}{8}$, $p_5 = \frac{11}{30}$, etc. These values are quite close together, and for large n they are all approximately .37.

To find the probability of having k matches, $0 \leq k \leq n$, is now quite easy. We let this probability be denoted p_n^k, and suppose $p_n^k = N_n^k/n!$ so that N_n^k is the number of orderings in which there are k matches. To construct the orderings that have precisely k matches, we could choose the positions at which to find matches, then for each such choice, order the remaining cards with no matches. For each of $\binom{n}{k}$ choices then there are N_{n-k}^0 ways to proceed. Thus, $N_n^k = \binom{n}{k} N_{n-k}^0$, and, dividing by $n!$, $p_n^k = p_{n-k}^0/k!$, where p_{n-k}^0 is found above. For example, the probability of 12 matches from 16 cards is $p_{16}^{12} = p_4^0/12! = 3/(8 \cdot 12!)$.

EXERCISES 3.5

1. Suppose a card is drawn from a standard deck of 52, noted and replaced, and another is drawn and noted. If all possible pairs of draws are taken to be equally likely, what is the probability of each of the following events?
- **(a)** Both cards drawn are red.
- **(b)** The first card drawn is red, and the second is black.
- **(c)** At least one card drawn is red.
- **(d)** One card drawn is an ace, and the other is a king.
- **(e)** The first card drawn is an ace, and exactly one card drawn is a diamond.

2. Suppose in Exercise 1 that the first card drawn is not replaced before the second is drawn. If all possible pairs of draws are taken to be equally likely, find the probability of each of the events listed in Exercise 1.

3. Suppose a 13-card hand is dealt from a deck of 52 and that all possible hands are deemed equally likely. What is the probability of each of the following events?

(a) The hand consists of 13 red cards.
(b) The hand consists of 13 hearts.
(c) The hand consists of seven hearts and six diamonds.
(d) The hand contains exactly seven hearts.
(e) The hand consists of four cards from one suit, three from each of the other three suits.
(f) The hand contains four aces.
(g) The hand contains exactly three aces.
(h) The hand contains no aces, kings, queens, jacks, or tens.

4. Suppose a five-card hand is dealt from a deck of 52 and that all possible hands are deemed equally likely. What is the probability of each of the following events?

(a) All cards are in one suit.
(b) The hand contains two aces and three kings.
(c) The hand contains two cards of one denomination and three of another.
(d) The five cards are in five different denominations.

5. A man chooses 10 numbers from 1, 2, 3, . . . , 80. Twenty balls are drawn from among 80 numbered 1, 2, 3, . . . , 80 in such a way that all draws are equally likely. What is the probability that none of the 20 numbers drawn match any of the ten chosen numbers? What is the probability that five of the 20 numbers drawn match five of the ten chosen numbers?

6. Suppose a die is thrown five times. Let Ω be the Cartesian product of the set $\{1, 2, 3, 4, 5, 6\}$ with itself five times, and let P be the equally likely measure on the events of Ω. What is the probability that

(a) Exactly three sixes are thrown?
(b) Exactly three sixes and exactly two fives are thrown?
(c) The sum of all throws is no greater than 6?
(d) All throws are different?

7. Ten prize-winning tickets are drawn from a bowl of t tickets, and all combinations of ten of these are regarded as equally likely. What is the probability that a man holding m tickets wins exactly two prizes? Wins at least one prize?

8. A man pulls two socks from a drawer in the dark. Suppose he has s socks in the drawer of which r are red and g are green ($g + r = s$), and suppose all combinations of two socks are equally likely to be drawn. What is the probability that he draws a pair of green socks? A matching pair of socks?

9. n married men at a square dance draw for partners from among the n wives in such a way that all possible matchings are equally likely. What is the probability that the tallest man is matched with his own wife? Is matched with the shortest woman? What is the probability that exactly four men are matched with their own wives?

10. Suppose w winning tickets are drawn from t tickets numbered $1, 2, \ldots, t$ in such a way that all combinations of w of these are equally likely. What is the probability that the highest numbered winning ticket is the one numbered r? What is the probability that the highest numbered winning ticket is the one numbered r and the lowest numbered winning ticket is the one numbered s?

11. A school has 40 freshmen, 30 sophomore, 20 junior, and 10 senior students. A committee of 10 students is selected in such a way that each combination of 10 students is equally likely to be selected. What is the probability that

(**a**) All committee members are freshmen?

(**b**) All committee members are freshmen or sophomores?

(**c**) The committee is composed of four freshmen, three sophomores, two juniors, and one senior?

(**d**) No committeemen are seniors?

12. Do Exercise 11 if the student body consists of 80 freshmen, 60 sophomores, 40 juniors, and 20 seniors.

13. See Example 3.8. What is the probability that Mr. Adams has four aces if he is playing poker with just Mr. Brown? What is the probability that he has four aces if he is playing poker with Mr. Brown, Mr. Clay, and Mr. Day?

14. Assuming the multiplication principle true for jobs of two stages, show that it is true for jobs of r stages (use induction).

15. How many eight-letter words may be formed from the letters of the word "happened"? How many four-letter words?

16. Suppose that the 52 cards of a standard deck are dealt 13 each to four players and that all deals of this sort are equally likely. What is the probability that

(**a**) One player receives all four aces?

(**b**) Every player receives 13 cards in one suit?

(**c**) Every player receives one ace?

17. The 52 cards of a standard deck are laid down successively from left to right in such a way that all orderings may be deemed equally likely. Find the probability of each of the following events.

(**a**) The 13 spades appear as the first 13 cards.

(**b**) The first spade to appear arrives as the sixth card.

(**c**) The ace of spades is the first card.

(**d**) The thirteenth heart to appear arrives before the thirteenth spade.

18. In how many ways may n people be seated around a table if two ways are regarded as the same whenever everyone has the same left- and right-hand neighbors? Suppose at a dinner party one of these arrangements is selected in such a way that all arrangements are equally likely. If a young man goes to the dinner party with his own true love, what is the probability that they are seated next to each other?

19. Suppose ten people are to be asked their birthdays and the 365^{10} possibilities are regarded as equally likely (February 29 is omitted). What is the probability that the ten people have ten different birthdays?

20. Suppose a coin is tossed n times. Let Ω be the Cartesian product of $\{H, T\}$ with itself n times, and let P be the equally likely measure on the events of Ω. What is the probability that exactly k of the tosses are heads?

21. A man requires his front door key and draws four keys from among the seven in his pocket in such a way that each combination of four is equally likely. If one of the seven keys is his front door key, what is the probability that he draws it? What is the answer to the same problem if he draws 40 keys from among 70?

22. Consider all possible eight-letter words formed from the alphabet of 26 letters to be equally likely. What is the probability that
(**a**) A word contains exactly two s's?
(**b**) A word contains at least one s?
(**c**) A word contains at least one s and exactly one r?
(**d**) A word contains at least one s and at least one r?
(**e**) A word begins with qxz?

23. n people of whom m are male are to be placed in a line from left to right in such a way that all permutations are equally likely. What is the probability that the oldest man is placed between two women? Is placed with a woman on his left?

24. A student takes a multiple-choice test composed of 100 questions each with six possible answers. If, for each question, he rolls a die to determine the answer to be marked, what is the probability that he scores 20 right?

25. Suppose k balls are successively placed in one of three boxes in such a way that all 3^k possible outcomes may be deemed equally likely. What is the probability that at least one box remains empty?

CHAPTER FOUR

Conditional Probability and Independence

4.1 Introduction

There are two important topics to be considered within the framework of a probability model consisting of a sample space Ω and a probability measure P on the events of Ω. The first is conditional probability; it arises from the need to adjust the probabilities of events once limited information about the outcome of the underlying experiment is available. As a part of this discussion, Bayes' theorem will be treated. The second topic is independence, a basic and peculiarly probabilistic concept. The first aspect of independence is the independence of events, the subject matter of Section 4.4. In Section 4.5, partitions of a sample space are introduced and independence in relation to partitions is treated.

4.2 Conditional probability

Consider a bridge player who cannot resist peeking at his cards one at a time as they are dealt, as opposed to picking up all 13 when the deal is complete. Ruling out superstition (for such a practice is viewed by some players as unlucky), we expect this player's 13-card hands to be probabilistically the same as those of another player who does not have this habit. That is, the probability that he picks up four aces or all red cards or any other set of hands is the same as it would otherwise be. However, as the cards are picked up and examined, the probabilities of various events concerning the particular 13-card hand that will be held on completion of the deal are influenced. Suppose, for example, that the ace of spades is the first card picked up. If the dealer were particularly sticky fingered, our bridge player might mentally make some of the following observations: this 13-card hand will contain at least one ace and at least one spade, it will not consist of all red cards, it appears that I have an improved chance of picking up all four aces, the remainder of my hand might be viewed as the selection of 12 cards from a deck of 51 with the ace of spades missing. After viewing 11 cards with two yet to come, he knows a great deal about the final composition of his hand. Some events may by now be precluded—for example, that of having all four aces if only one has shown up so far. Some events may now be certain—for example, that of having four aces if they are among the 11 cards already received. Knowing the placement of 11 cards in the deal, the player may view the two to come as the selection of two cards from a deal of 41 missing the 11 in hand and thereby make certain probability calculations concerning the entire hand. We intend to exploit just this kind of observation in using information received.

There are a variety of complex experiments in which the final outcome is determined in a sequence of stages. In such experiments, like the one described above, it is necessary to consider adjusted probability measures that take into account the information made available by the outcomes of the early stages. In the same vein, imagine a husband and wife who decide to flip a coin seven times to settle an argument. They agree to go out to dinner if more heads appear than tails and to eat at home if more tails appear than heads. If all 2^7 sequences of heads and tails of length seven are considered equally likely, the probability of going out to dinner is $\frac{1}{2}$. Suppose, however, that the first toss is a tail. One then feels that the probability that they will go out to dinner has diminished. Our answer would be that it has, and, specifically, that it is now only $\frac{11}{32}$. Again, if the first four tosses have produced one head and three tails, one expects a dinner at home. Given this information, we will find the probability of going out to dinner to be only $\frac{1}{8}$. Of course, if five tosses have proved one head and four tails, the coin flipping would usually be abandoned as a waste of effort, eating at home being certain.

Suppose Ω is a sample space. Partial information about the outcome $\omega \in \Omega$ like that above may be abstracted in a particularly simple form. It is accounted for by saying that some event A from Ω has occurred. Thus, for the first example, if Ω consists of descriptions of all 13-card hands from a deck of 52 and if the ace of spades is drawn first, we know that $A = \{\omega | \ \omega \text{ has in it the ace of spades}\}$ has occurred. That is, whatever the final outcome ω, it is a member of A. If 11 cards are drawn and noted and two more remain to be drawn, then $A = \{\omega | \ \omega \text{ has in it the 11 cards noted}\}$ has occurred. In the second example, if Ω consists of all sequences of heads and tails of length seven and if the first toss produces a tail, we are aware of the fact that $A = \{\omega | \ \omega$ begins with a tail$\}$ has occurred. If the first four tosses produce one head and three tails, then we know that $A = \{\omega | \ \omega$ begins with one head and three tails in some order$\}$ has occurred. In all these cases, the outcome ω has been positioned in some proper subset of Ω.

We now consider the influence on probabilities of the knowledge that some event has occurred. Let Ω be a fixed sample space, and let P be a probability measure on the events of Ω. Suppose it is known that the event A from Ω has occurred where $P(A) > 0$.

Definition 4.1 For any event B from Ω, the *conditional probability* of B given A, written $P_A(B)$, is defined by $P_A(B) = P(A \cap B)/P(A)$.

Therefore, P_A is a new function of the events of Ω defined through use of the measure P.

The computation of conditional probabilities hinges on the computation of ordinary probabilities. In particular, if P is equally likely on Ω, $P(A \cap B) = N(A \cap B)/N(\Omega)$ (N again means the number of elements in) and $P(A) = N(A)/N(\Omega)$. Then

$$P_A(B) = \frac{N(A \cap B)}{N(\Omega)} \cdot \frac{N(\Omega)}{N(A)} = \frac{N(A \cap B)}{N(A)}$$

so that here we do not even need to find $N(\Omega)$. Before any conditional probabilities are explicitly found, let us make a small investigation of what we have done in defining conditional probability given an event A.

The main point we wish to emphasize is that P_A is itself a probability measure on the events of Ω. Returning to the definition of a probability measure, we find that P_A must satisfy

(1) $P_A(B) \geq 0$ for any event B from Ω.

(2) $P_A(\Omega) = 1$.

(3) $P_A(B \cup C) = P_A(B) + P_A(C)$ for any disjoint events from Ω.

Are these conditions satisfied? Condition (1) depends only on the fact that $P(A \cap B) \geq 0$ and $P(A) > 0$. Condition (2) follows from noting that $A \cap \Omega = A$ and therefore that

$$P_A(\Omega) = \frac{P(A \cap \Omega)}{P(A)} = \frac{P(A)}{P(A)} = 1$$

Suppose then that B and C are disjoint events from Ω. This implies that $A \cap B$ and $A \cap C$ are also disjoint and that $P(A \cap (B \cup C)) = P((A \cap B) \cup (A \cap C)) = P(A \cap B) + P(A \cap C)$. Consequently,

$$P_A(B \cup C) = \frac{P(A \cap (B \cup C))}{P(A)} = \frac{P(A \cap B) + P(A \cap C)}{P(A)}$$
$$= P_A(B) + P_A(C)$$

A probability measure on the events of Ω is determined by its values at the simple events of Ω. Consider then $P_A(\{\omega\})$ for $\omega \in \Omega$. Two cases are appropriate: either $A \cap \{\omega\} = \{\omega\}$ or $A \cap \{\omega\} = \varnothing$. The first of these happens if $\omega \in A$, the second if $\omega \notin A$. Respectively, then,

$$P_A(\{\omega\}) = \frac{P(\{\omega\})}{P(A)} \quad \text{and} \quad P_A(\{\omega\}) = \frac{P(\varnothing)}{P(A)} = 0$$

This may be summarized as follows: given A, the simple events $\{\omega\}$ with $\omega \in A$ have conditional probabilities that are their original probabilities scaled upwards by a factor $1/P(A)$ $(1/P(A) \geq 1)$; the simple events $\{\omega\}$ with $\omega \notin A$ have conditional probabilities given A equal to zero.

The probability measure P_A has the following intuitive description. For any event B, $P_A(B)$ measures the probability of occurrence of B, conditioned by the knowledge that the event A has occurred. It can be verified that this definition of P_A has a number of properties that this interpretation suggests. For example, $P_A(A) = 1$, meaning that the probability at A will occur given A, is one. Similarly, given A, it is impossible that A' will occur. This is the interpretation of the formal fact that $P_A(A') = 0$.

Again, it is clear from the definition that $P(A \cap B) = P(A)P_A(B)$. This equation is often read as follows: the probability of the joint occurrence of A and B is equal to the probability of A multiplied by the conditional probability of B given A. This phrasing has an obvious intuitive appeal and is indeed partly the motivation for Definition 4.1.

Example 4.1

The Peeking Bridge Player Let $A = \{\omega |\ \omega$ has in it the ace of spades$\}$, and suppose P is equally likely over a sample space of descriptions of all possible 13-card hands from a deck of 52. What is the conditional probability of the event B of having all four aces given the event A?

In line with the remark made above, it suffices in this equally likely case to determine $N(A \cap B)$ and $N(A)$. $N(A)$ is clearly the number of 12-card hands which may be formed from 51 cards, $\binom{51}{12}$. Moreover, $B \subset A$, so that $N(A \cap B) = N(B)$ is the number of hands that contain all four aces, $\binom{48}{9}$. The required conditional probability is

$$P_A(B) = \frac{\binom{48}{9}}{\binom{51}{12}} = \frac{48!}{39!9!}\,\frac{39!12!}{51!} = \frac{12 \cdot 11 \cdot 10}{51 \cdot 50 \cdot 49}$$

That this is in fact larger than

$$P(B) = \frac{\binom{48}{9}}{\binom{52}{13}} = \frac{13 \cdot 12 \cdot 11 \cdot 10}{52 \cdot 51 \cdot 50 \cdot 49}$$

is immediately apparent—it is $1/P(A) = 4$ times as large. Next consider the conditional probability of the event C that the hand contains four kings given A. Here $A \cap C$ is the event that the hand contains four kings and the ace of spades. $N(A \cap C)$ is therefore $\binom{47}{8}$, and the conditional probability $P_A(C)$ is

$$\frac{\binom{47}{8}}{\binom{51}{12}} = \frac{12 \cdot 11 \cdot 10 \cdot 9}{51 \cdot 50 \cdot 49 \cdot 48}$$

which is slightly smaller than

$$P(C) = \frac{13 \cdot 12 \cdot 11 \cdot 10}{52 \cdot 51 \cdot 50 \cdot 49}$$

in fact $\frac{3}{4}$ as large. We might "explain" this by the observation that a hand known to have the ace of spades as one card has less room for kings. This phenomenon will prove not uncommon.

Example 4.2

The Dinner Argument In the case of the husband and wife who cannot agree on a dining arrangement, we are adopting a sample space Ω of all sequences of heads and tails of length seven and an equally likely probability measure on Ω. The probability of a majority of heads is clearly $\frac{1}{2}$. Now, suppose the first flip results in a tail. What is the conditional probability that the couple will go out to dinner, given that the first toss resulted in a tail? The appropriate sequences are now all 2^6 sequences of heads and tails of length seven that begin with tails. A sequence from among these that is compatible with going out to dinner must have at least four heads in the last six positions, i.e., four heads or five heads or six heads. There are, respectively, $\binom{6}{4} = 15$, $\binom{6}{5} = 6$, and $\binom{6}{6} = 1$ of these sequences. The conditional probability in question is therefore

$$\frac{N(\{\omega \mid \omega \text{ begins with tails and has a majority of heads}\})}{N(\{\omega \mid \omega \text{ begins with tails}\})} = \frac{22}{64} = \frac{11}{32}$$

If the first four tosses have resulted in one head and three tails, the appropriate sequences are the $2^3 = 8$ that begin TTTH, the eight that begin TTHT, the eight that begin THTT, and the eight that begin HTTT. In each group of eight there is precisely one sequence that allows going out to dinner, for it is clear that three heads must be found to obtain a majority. Hence, the conditional probability of going out to dinner here is $\frac{4}{32} = \frac{1}{8}$.

It is worth observing that these answers could have as well been found by ignoring the first tosses and dealing only with sequences of length six and length three, respectively. Of course, to obtain a majority of heads overall, one needs to find enough heads in these shorter sequences. In either case they could be viewed as occupancy problems in which tosses are successively placed in two boxes labelled heads and tails. In the first case, six tosses are to be placed in such a way that at least four go into the heads box; in the second case, three tosses are to be placed all in the heads box. In such light, the answers found above could be obtained a little more directly from the answers found in the occupancy example of Section 3.5 (Example 3.9).

EXERCISES 4.2

1. Let P be the equally likely probability measure on the events of $\Omega = \{\omega_1, \omega_2, \omega_3, \omega_4, \omega_5, \omega_6\}$. Determine the conditional probability of
(a) $\{\omega_1, \omega_3, \omega_4\}$ given $\{\omega_2, \omega_3\}$.
(b) $\{\omega_2, \omega_3\}$ given $\{\omega_1, \omega_3, \omega_4\}$.
(c) $\{\omega_1, \omega_2, \omega_3, \omega_4\}$ given $\{\omega_3, \omega_4, \omega_5, \omega_6\}$.
(d) $\{\omega_1, \omega_2, \omega_3, \omega_4\}$ given $\{\omega_2, \omega_3\}$.
(e) $\{\omega_2, \omega_3\}$ given $\{\omega_1, \omega_2, \omega_3\}$.

2. Do (a) through (e) of Exercise 1, if P is a probability measure determined on the events of Ω by $P(\{\omega_1\}) = \frac{1}{8}$, $P(\{\omega_2\}) = \frac{1}{8}$, $P(\{\omega_3\}) = \frac{1}{3}$, $P(\{\omega_4\}) = \frac{1}{12}$, $P(\{\omega_5\}) = \frac{1}{6}$, $P(\{\omega_6\}) = \frac{1}{6}$.

3. For the peeking bridge player, find the conditional probability of
(a) Having exactly four spades given that he has the ace of spades.
(b) Having exactly three hearts given that he has the ace of spades.
(c) Having at least two aces given that he has the ace of spades.
(d) Having four diamonds given that he has the ace of spades, the king of hearts, and the queen of clubs.

4. Let Ω be a sample space of descriptions of all thirteen card hands from an ordinary deck of 52, and let P be equally likely on the events of Ω. Let $A = \{\omega|\ \omega$ has in it the ace of spades$\}$, $B = \{\omega|\ \omega$ has in it at least one ace$\}$, and $C = \{\omega|\ \omega$ has in it at least two aces$\}$. Determine $P_A(B)$, $P_A(C)$, $P_B(A)$, $P_B(C)$, $P_C(A)$, and $P_C(B)$.

5. Suppose r men are drafted among m eligible men in such a way that each combination of r men has the same chance of being chosen. What is the probability that the oldest eligible man is drafted? What is the conditional probability that the oldest eligible man is drafted given that the youngest eligible man is drafted? Is not drafted?

6. A red and a green die are thrown, and P is supposed equally likely on the events of the sample space $\Omega = \{(\text{R1, G1}), (\text{R2, G1}), \ldots, (\text{R6, G1}), (\text{R1, G2}), \ldots, (\text{R6, G6})\}$ of 36 sample points. What is the conditional probability of
(a) A sum of faces of seven given that the red die is odd.
(b) The red die is odd given a sum of faces of seven.
(c) A sum of faces of seven given that the red die is two.

(**d**) The red die is two given a sum of faces of seven.
(**e**) A sum of faces of seven given that both dice are odd.
(**f**) A sum of faces of at least seven given that the face on the red die minus the face on the green die is one.
(**g**) A sum of faces of at least seven given that the red die is at least three.
(**h**) The red die is at least three given a sum of faces of at least seven.

7. For the dinner argument, find the conditional probability of going out to dinner given that the first two tosses are tails and given that the first three tosses produce one head and two tails.

8. Suppose n husbands and their wives at a square dance are successively paired off by having the husbands draw names from a hat. What is the probability that no husband draws his own wife given that the first one does not draw his?

9. A man throws a die three times in succession. Suppose P is equally likely on the sample space of sequences of 1, 2, 3, 4, 5, 6 of length three. What is the conditional probability of
(**a**) A sum on three tosses of at least ten given the first toss is two.
(**b**) Three tosses of six given at least two tosses of six.
(**c**) Three tosses the same given the first toss is two.
(**d**) A sum of three tosses of 16 given at least two tosses the same.

10. Suppose that P is a probability measure on the events of a sample space Ω and that $\{\omega\}$ is a simple event from Ω with $P(\{\omega\}) > 0$. Determine $P_{\{\omega\}}(A)$ for any event A from Ω. Interpret the answer.

11. Show that if P is a probability measure on the events of a sample space Ω, then $P_A(A) = 1$ and $P_A(A') = 0$ for any event A of Ω with $P(A) > 0$.

12. Suppose that P is a probability measure on the events of a sample space Ω. Show that if A, B, and C are events from Ω with $B \subset A$, $C \subset A$, and $P(C) > 0$, then $P(B)/P(C) = P_A(B)/P_A(C)$.

13. Let P be a probability measure on the events of a sample space Ω, and let $A \subset B$ be events of Ω with $P(A) > 0$ and $P(B) > 0$. Show that $P(A) \leq P_B(A)$ and that $P_A(B) = 1$.

14. Suppose that P is a probability measure on the events of a sample space Ω and that A is an event from Ω with $P(A) = 1$. Show that $P_A(B) = P(B)$ for all $B \subset \Omega$. Explain.

15. Let P be a probability measure on the events of sample space Ω, and let A and B be two events from Ω with $P(B) > 0$ and $0 < P(A) < 1$. Show that
(**a**) $P_A(B) > P(B)$ implies $P_B(A) > P(A)$ and $P_{A'}(B) < P(B)$.
(**b**) $P_A(B) = P(B)$ implies $P_B(A) = P(A)$ and $P_{A'}(B) = P(B)$.

16. For the occupancy example of Section 3.5 (Example 3.9), what is the conditional probability of finding k_2 balls in the second box given k_1 in the first box? What is the conditional probability of finding k_2 balls in the second box, k_3 in the third box, $\ldots$, k_q in the qth box given k_1 in the first box?

4.3 Bayes' theorem

Suppose a sports fan on a world tour hears, by way of overseas radio, that the New York Mets have defeated the Los Angeles Dodgers in a baseball game. If he knows the probability of this occurrence is $\frac{1}{4}$ when the game is played in New York and $\frac{1}{10}$ when the game is played in Los Angeles, can a probability be attached to the event that the game was played in New York? That is, does it make sense to talk of the conditional probability of a New York location given a Mets victory?

In order to give a positive answer to this question, another ingredient needs to be added—the unconditional probability of a New York location. To see why this is an important ingredient, we consider three possible unconditional probabilities of a New York location for three different situations. From a frequency point of view one might set this probability at $\frac{1}{2}$ if no further information is available. This will reflect the fact that the total number of games played between the two teams is always divided equally between the two home locations. If this is done, we shall find below that the conditional probability of a New York location given a Mets victory is $\frac{5}{7}$. Here the probability of a New York location has been enhanced by conditioning on the event of a Mets victory. Suppose rather that the fan knows that at this particular time of the year, games between the two teams are more frequently played in Los Angeles than in New York, say, $\frac{3}{4}$ of them. The conditional probability in question will now be found to be $\frac{5}{11}$, again slightly higher than the unconditional probability of $\frac{1}{4}$. Finally, suppose the fan has a schedule that informs him that the game was played in Los Angeles. The conditional probability of the game having been played in New York given a Mets victory must in fact be zero.

The calculations being performed in these instances come from the following statement within our mathematical model. Suppose Ω is a sample space and P is a probability measure on the events of Ω. Let $A, B_1, B_2, \ldots, B_m$ be events from Ω, all with positive probability.

Bayes' Theorem If $B_1, B_2, \ldots, B_m$ are mutually exclusive and exhaustive, then

$$P_A(B_i) = \frac{P(B_i)P_{B_i}(A)}{\sum_{j=1}^{m} P(B_j)P_{B_j}(A)}, \qquad i = 1, 2, \ldots, m$$

PROOF: The purported numerator of $P_A(B_i)$ is

$$P(B_i)P_{B_i}(A) = P(B_i)\,\frac{P(A \cap B_i)}{P(B_i)} = P(A \cap B_i)$$

The purported denominator of $P_A(B_i)$ is

$$\sum_{j=1}^{m} P(B_j)P_{B_j}(A) = \sum_{j=1}^{m} P(A \cap B_j) = P(\bigcup_{j=1}^{m} (A \cap B_j))$$

$$= P(A \cap \Omega) = P(A)$$

The result follows from the definition of $P_A(B_i)$.

This theorem is most often viewed as a formula for $P_A(B_i)$, $i = 1, 2, \ldots, m$, when the probabilities $P(B_i)$ and $P_{B_i}(A)$, $i = 1, 2, \ldots, m$, are at hand.

How does this all reflect on the Dodgers and the Mets? In the first place, we know that to be able to answer questions of the kind asked we must keep track of both the location of the game and the winning team in tabulating outcomes. A suitable sample space might then be $\Omega = \{(\text{NY, M}), (\text{NY, D}), (\text{LA, M}), (\text{LA, D})\}$, where (NY, M) corresponds to a New York location and a Mets victory, etc. The events $A = \{(\text{NY, M}), (\text{LA, M})\}$, $B_1 = \{(\text{NY, M}), (\text{NY, D})\}$, and $B_2 = \{(\text{LA, M}), (\text{LA, D})\}$ are the events of a Mets victory, of a New York location, and of a Los Angeles location, respectively. B_1 and B_2 are mutually exclusive and exhaustive. The original assumptions made take the form $P_{B_1}(A) = \frac{1}{4}$ and $P_{B_2}(A) = \frac{1}{10}$. If now $P(B_1)$ and $P(B_2)$ are each taken to be $\frac{1}{2}$, Bayes' theorem may be invoked, and we find

$$P_A(B_1) = \frac{\frac{1}{2}\cdot\frac{1}{4}}{\frac{1}{2}\cdot\frac{1}{4} + \frac{1}{2}\cdot\frac{1}{10}} = \frac{5}{7}, \qquad P_A(B_2) = \frac{\frac{1}{2}\cdot\frac{1}{10}}{\frac{1}{2}\cdot\frac{1}{4} + \frac{1}{2}\cdot\frac{1}{10}} = \frac{2}{7}$$

The second choice was $P(B_1) = \frac{1}{4}$ and therefore $P(B_2) = \frac{3}{4}$. Then

$$P_A(B_1) = \frac{\frac{1}{4}\cdot\frac{1}{4}}{\frac{1}{4}\cdot\frac{1}{4} + \frac{3}{4}\cdot\frac{1}{10}} = \frac{5}{11}, \qquad P_A(B_2) = \frac{\frac{3}{4}\cdot\frac{1}{10}}{\frac{1}{4}\cdot\frac{1}{4} + \frac{3}{4}\cdot\frac{1}{10}} = \frac{6}{11}$$

In the last example, $P(B_1) = 0$ and $P(B_2) = 1$, but then $P_A(B_1) = 0$ and $P_A(B_2) = 1$.

In each of these cases, the assumptions made have allowed us to find a probability measure P on all the events of Ω. To illustrate, let us consider just the choice of $P(B_1) = P(B_2) = \frac{1}{2}$. The simple event $\{(\text{NY, M})\}$ is recognizable as $A \cap B_1$ and therefore $P(\{\text{NY, M}\}) = P(A \cap B_1) = P(B_1)P_{B_1}(A) = \frac{1}{2}\cdot\frac{1}{4} = \frac{1}{8}$. In like manner, $\{(\text{NY, D})\}$ is the event $A' \cap B_1$ and $P(\{(\text{NY, D})\}) = P(A' \cap B_1) = P(B_1)P_{B_1}(A') = \frac{1}{2}\cdot\frac{3}{4} = \frac{3}{8}$. Continuing in this way, $P(\{(\text{LA, M})\}) = \frac{1}{20}$ and $P(\{(\text{LA, D})\}) = \frac{9}{20}$. Once this has all been fixed, it is simple matter to compute $P(A) = P(\{(\text{NY, M}), (\text{LA, M})\}) = \frac{1}{8} + \frac{1}{20} = \frac{7}{40}$ and then

$$P_A(B_1) = \frac{P(A \cap B_1)}{P(A)} = \frac{\frac{1}{8}}{\frac{7}{40}} = \frac{5}{7}$$

In the statement of Bayes' theorem we should stress one particular aspect. Since the events $B_1, B_2, \ldots, B_m$ are mutually exclusive and exhaustive,

$$P(\Omega) = P(\bigcup_{i=1}^{m} B_i) = \sum_{i=1}^{m} P(B_i) = 1$$

Similarly,

$$P_A(\Omega) = P_A(\bigcup_{i=1}^{m} B_i) = \sum_{i=1}^{m} P_A(B_i) = 1$$

The probabilities $P(B_1), P(B_2), \ldots, P(B_m)$ are called the *a priori* probabilities of the events $B_1, B_2, \ldots, B_m$ while the (conditional) probabilities $P_A(B_1), P_A(B_2), \ldots, P_A(B_m)$ are called their *a posteriori* probabilities. Bayes' theorem provides the transition from a priori to a posteriori (after the occurrence of A) probabilities of mutually exclusive and exhaustive events. In the examples above, these transitions are from $\frac{1}{2}, \frac{1}{2}$ to $\frac{5}{7}, \frac{2}{7}$, from $\frac{1}{4}, \frac{3}{4}$ to $\frac{5}{11}, \frac{6}{11}$, and from 0, 1 to 0, 1.

Example 4.3

Suppose three urns numbered 1, 2, and 3 contain, respectively, one red and one black ball, two red and three black balls, and four red and two black balls. Consider an experiment consisting of the selection of an urn followed by the draw of a ball from it. For a sample space, we might take $\Omega = \{(1, \text{R}), (1, \text{B}), (2, \text{R}), (2, \text{B}), (3, \text{R}), (3, \text{B})\}$ with an obvious interpretation of sample points. $B_1 = \{(1, \text{R}), (1, \text{B})\}$, $B_2 = \{(2, \text{R}), (2, \text{B})\}$, and $B_3 = \{(3, \text{R}), (3, \text{B})\}$ are the events described by the selection of urn 1, 2, and 3, respectively. They are, moreover, mutually exclusive and exhaustive. If $A = \{(1, \text{R}), (2, \text{R}), (3, \text{R})\}$ and we suppose each ball in the selected urn is equally likely to be drawn, then $P_{B_1}(A) = \frac{1}{2}$, $P_{B_2}(A) = \frac{2}{5}$, and $P_{B_3}(A) = \frac{2}{3}$. If the urn number is not observed but a red ball is drawn, what is the probability that it was drawn from urn 1, urn 2, or urn 3? That is, what are the a posteriori probabilities $P_A(B_1)$, $P_A(B_2)$, and $P_A(B_3)$? The answer to this depends on the a priori probabilities $P(B_1)$, $P(B_2)$, and $P(B_3)$.

Let us suppose, for this model, that $P(B_1) = P(B_2) = P(B_3) = \frac{1}{3}$. Using Bayes' theorem, we find

$$P_A(B_1) = \frac{\frac{1}{3}\cdot\frac{1}{2}}{\frac{1}{3}\cdot\frac{1}{2} + \frac{1}{3}\cdot\frac{2}{5} + \frac{1}{3}\cdot\frac{2}{3}} = \frac{15}{47}$$

$$P_A(B_2) = \frac{\frac{1}{3}\cdot\frac{2}{5}}{\frac{1}{3}\cdot\frac{1}{2} + \frac{1}{3}\cdot\frac{2}{5} + \frac{1}{3}\cdot\frac{2}{3}} = \frac{12}{47}$$

$$P_A(B_3) = \frac{\frac{1}{3}\cdot\frac{2}{3}}{\frac{1}{3}\cdot\frac{1}{2} + \frac{1}{3}\cdot\frac{2}{5} + \frac{1}{3}\cdot\frac{2}{3}} = \frac{20}{47}$$

Here the probabilities of urn 1 and urn 2 have lessened while that of urn 3 has increased due to its composition of balls. If rather the a priori probabilities are $P(B_1) = \frac{1}{2}$, $P(B_2) = \frac{1}{3}$, and $P(B_3) = \frac{1}{6}$, we find

$$P_A(B_1) = \frac{\frac{1}{2}\cdot\frac{1}{2}}{\frac{1}{2}\cdot\frac{1}{2} + \frac{1}{3}\cdot\frac{2}{5} + \frac{1}{6}\cdot\frac{2}{3}} = \frac{45}{89}$$

$$P_A(B_2) = \frac{\frac{1}{3}\cdot\frac{2}{5}}{\frac{1}{2}\cdot\frac{1}{2} + \frac{1}{3}\cdot\frac{2}{5} + \frac{1}{6}\cdot\frac{2}{3}} = \frac{24}{89}$$

$$P_A(B_3) = \frac{\frac{1}{6}\cdot\frac{2}{3}}{\frac{1}{2}\cdot\frac{1}{2} + \frac{1}{3}\cdot\frac{2}{5} + \frac{1}{6}\cdot\frac{2}{3}} = \frac{20}{89}$$

Again $P_A(B_3) > P(B_3)$, but here the a posteriori probability of B_3 is still the smallest since its a priori probability was small.

The language of ball and urn may seem peculiar. For problems of the kind discussed in this section, however, we might imagine the events B_i, $i = 1, 2, \ldots, m$ to correspond to the selection of urns numbered $1, 2, \ldots, m$ and give each urn an appropriate composition of red and black balls to reflect the conditional probability $P_{B_i}(A)$. Thus, the model is quite general and is furthermore a handy memory device.

Remarks: There is a certain amount of persistent confusion surrounding Bayes' theorem. The theorem itself represents only an easy manipulation of probabilities, and, within the mathematical framework of a probability measure on the events of a sample space, it is not subject to dispute. Difference of opinion does exist (even among "experts") as to which experimental situations allow its proper use. This difference is rooted in the interpretation of "The probability of an event A is $P(A)$" as discussed in Section 3.3. It ordinarily centers about the existence and/or meaning of the a priori probabilities as set out above. Thus, in the urn model, should one assign probabilities to the selection of an urn? The answer to this question will depend again on the nature of the underlying experiment and the person empowered to answer. We have attempted to stress the fact that when the answer is yes, the a posteriori probabilities depend rather heavily on the choice of a priori probabilities.

EXERCISES 4.3

1. Suppose $P_{B_1}(A) = \frac{1}{4}$, $P_{B_2}(A) = \frac{1}{2}$, $P_{B_3}(A) = \frac{3}{4}$, and $P_{B_4}(A) = 1$. Find the a posteriori probabilities of B_1, B_2, B_3, and B_4 given A if the a priori probabilities are

(**a**) $P(B_1) = P(B_2) = P(B_3) = P(B_4) = \frac{1}{4}$.
(**b**) $P(B_1) = P(B_2) = \frac{1}{8}$, $P(B_3) = P(B_4) = \frac{3}{8}$.
(**c**) $P(B_1) = \frac{2}{5}$, $P(B_2) = \frac{3}{10}$, $P(B_3) = \frac{1}{5}$, $P(B_4) = \frac{1}{10}$.

2. For (**a**), (**b**), and (**c**) of Exercise 1, find the a posteriori probabilities of B_1, B_2, B_3, and B_4 given A'.

3. Let Ω be a set of descriptions of all five-card hands from a deck of 52, and let P be equally likely on the events of Ω. Suppose $B_i = \{\omega | \omega \text{ has in it } i \text{ aces}\}$, $i = 0, 1, 2, 3, 4$, and $A = \{\omega | \omega \text{ has in it three black cards}\}$. Find the a priori probabilities of B_0, B_1, B_2, B_3, and B_4 and their a posteriori probabilities given A. For which of B_0, B_1, B_2, B_3, and B_4 is the a posteriori probability higher than the a priori probability?

4. A student taking a true-false test always marks the correct answer when he knows it and decides true or false on the basis of flipping a coin when he does not know it. If the probability that he will know an answer is $\frac{3}{5}$, what is the probability that he knew the answer to a correctly marked question?

5. A college entrance board views students as capable or incapable. The examination given to prospective students is passed by a capable student with probability .9 and with probability .3 by an incapable student. If the probability that a student taking the exam is capable is .45, what is the probability that a student who has passed is capable?

6. A basketball recruiter observes that 50% of all good players stand in excess of 7 feet while only 10% of the poor players do. If 30% of all players are good, what percentage of players who stand in excess of 7 feet are good? What is the percentage of players who stand this tall?

7. Prisoners Adams, Brown, and Clay are informed that their names have been placed in a hat and that one name has been drawn. The prisoner whose name has been drawn will be put to death while the other two will be allowed to go free. Adams asks the messenger of these tidings to name either Brown or Clay as one of those who will go free since at least one of them will. The messenger refuses on the basis that Adams would thereby increase his probability of extinction from $\frac{1}{3}$ to $\frac{1}{2}$. What is wrong? [Begin by setting up a sample space Ω of pairs like (b, c) to represent the outcome that the messenger says Brown goes free and Clay's name has been drawn.]

4.4 Independence of events

The concept of independence will have, for us, a variety of forms. This section is devoted to the independence of events upon which all succeeding forms are based. Each version of independence brought forward will intuitively mean freedom from influence in an appropriate sense; thus, the choice of terminology has a descriptive merit.

Suppose Ω is a sample space and P is a probability measure on the events of Ω. Let A and B be two events from Ω each with positive probability. Let us suppose that $P_A(B) = P(B)$, i.e., the conditional probability of B given A is no different than its original probability. We might view this as the statement that the occurrence of A has no influence on the probability of the occurrence of B. This relationship might be contrasted with $P_A(B) > P(B)$, say, in which case the occurrence of A enhances the probability of occurrence of B. Now, in order that $P_A(B) = P(B)$, it must be that $P(A \cap B) = P(A)P(B)$. Written in this form, it is apparent that $P_B(A) = P(A)$ also, i.e., there is no influence of the occurrence of B on the probability of the occurrence of A either. We choose to call this property the independence of A and B, and to single it out we use the symmetric condition $P(A \cap B) = P(A)P(B)$.

Definition 4.2 Two events A and B from Ω are called *independent* events if $P(A \cap B) = P(A)P(B)$; they are called *dependent* events if $P(A \cap B) \neq P(A)P(B)$.

It may be remarked immediately that Ω and A are independent events for any $A \subset \Omega$. This fact follows on noting that $P(A \cap \Omega) = P(A) = P(A)P(\Omega)$. In a similar way, one finds that $\varnothing$ and A are independent events for any event A from Ω (this in spite of the fact that conditional probability given $\varnothing$ is not defined).

Before we study some concrete examples, we can exploit the intuitive notion of absence of influence or, more especially, the lack thereof in a general setting. Suppose for example, that A and B are disjoint events. If so, the occurrence of A implies the nonoccurrence of B, and independence of these events should *not* be expected. Formally, $P(A \cap B) = P(\varnothing) = 0 \neq P(A)P(B)$ unless one or both of $P(A)$ and $P(B)$ are zero. In the same vein, if $A \subset B$, the occurrence of A would imply the occurrence of B so one usually expects the *dependence* of A and B. Here $P(A \cap B) = P(A) \neq P(A)P(B)$ unless $P(B) = 1$ or $P(A) = 0$.

Example 4.4

Let us take the experiment of selecting a 13-card hand from a deck of 52. Let Ω be a sample space of descriptions of all possible 13-card hands, and let P be equally likely on Ω. It is easy to locate events A and B from Ω which are dependent—we need only use the device above. For example, take $A = \{\omega | \ \omega$ has in it the ace of spades$\}$, and take $B = A'$; then A and B dependent. If rather $B = \{\omega | \ \omega$ has in it all four aces$\}$, then $B \subset A$, and they are again dependent events.

To find examples of independent events, we must avoid the obviously dependent events like those above. Let us consider the events A above and the events $B = \{\omega | \ \omega$ has in it the three of clubs$\}$. The computation below shows that these are actually dependent events. Although it is not as obvious in the present case as it was in that above, nevertheless they are dependent events. Specifically,

$$P(A \cap B) = \frac{\binom{50}{11}}{\binom{52}{13}} = \frac{13 \cdot 12}{52 \cdot 51} = \frac{1}{17} \neq \frac{1}{16} = \frac{\binom{51}{12}\binom{51}{12}}{\binom{52}{13}\binom{52}{13}} = P(A)P(B)$$

Thus, the conditional probability of having the three of clubs given the ace of spades is $\frac{4}{17}$ while the probability of having the three of clubs is $\frac{1}{4} > \frac{4}{17}$.

Let us try the same event A and now let $B = \{\omega | \ \omega$ has in it 13 cards in the same suit$\}$. We find

$$P(A \cap B) = \frac{1}{\binom{52}{13}} = \frac{1}{4} \cdot \frac{4}{\binom{52}{13}} = P(A)P(B)$$

so these events are independent. This is explicable to the extent that a one-suited hand must be in some suit and we should not be able to rule it out or consider its probability of occurrence enhanced or lessened by fixing one card. As a second example, consider $A = \{\omega | \ \omega$ has in it the aces of spades and hearts$\}$ and $B = \{\omega | \ \omega$ has in it more red cards than black$\}$. Clearly, $P(B) = \frac{1}{2}$. What about $P_A(B)$? We claim $P_A(B) = \frac{1}{2}$ also so that A and B are independent events.

An important feature of independence is that the independence of some events will imply the independence of other events. For example, if A and B are independent events from Ω, then so are A and B'. To demonstrate this, we must show that the relation $P(A \cap B') = P(A)P(B')$ follows from the relation $P(A \cap B) = P(A)P(B)$. To see that this implication holds, we write first of all $P(A) = P(A \cap B) + P(A \cap B') = P(A)P(B) + P(A \cap B')$, then, $P(A \cap B') = P(A) - P(A)P(B) = P(A)(1 - P(B)) = P(A)P(B')$. Reversing the roles of A and B will show A' and B are independent events when A and B are. The argument above applied to A' and B then shows that A' and B' are independent events when A and B are. Here the independence of three other pairs of events follows from the independence of just one pair of events.

What will be called the *mutual* independence of more than two events is not directly suggested by Definition 4.2. Let $A_1, A_2, \ldots, A_k$ be events from a sample space Ω on which P is a probability measure.

Definition 4.3 $A_1, A_2, \ldots, A_k$ are said to be *mutually independent* if for every selection of two or more events from $A_1, A_2, \ldots, A_k$, say $A_{i_1}, A_{i_2}, \ldots, A_{i_r}$, we have $P(\bigcap_{j=1}^{r} A_{i_j}) = \prod_{j=1}^{r} P(A_{i_j})$.

First, let us compare this definition with Definition 4.2 when $k = 2$. The only selection of two or more events from A_1 and A_2 is that of A_1 and

A_2. For mutual independence of A_1 and A_2 it must be that $P(A_1 \cap A_2) = P(A_1)P(A_2)$ or, in other words, A_1 and A_2 must be independent events. The two definitions given for $k = 2$ then agree. The conditions that must be satisfied for the mutual independence of three events A_1, A_2, and A_3 are

(1) $P(A_1 \cap A_2) = P(A_1)P(A_2)$.
(2) $P(A_1 \cap A_3) = P(A_1)P(A_3)$.
(3) $P(A_2 \cap A_3) = P(A_2)P(A_3)$.
(4) $P(A_1 \cap A_2 \cap A_3) = P(A_1)P(A_2)P(A_3)$.

The first three conditions are that A_1 and A_2 be independent, that A_1 and A_3 be independent, and that A_2 and A_3 be independent. Condition (4) does not appear to deal directly with the independence of a pair of events in the way that (1), (2), and (3) do. However, (1) and (4) taken together show that

$$P((A_1 \cap A_2) \cap A_3) = P(A_1 \cap A_2 \cap A_3) = P(A_1)P(A_2)P(A_3) = P(A_1 \cap A_2)P(A_3)$$

so the mutual independence of A_1, A_2, and A_3 implies the independence of the events $A_1 \cap A_2$ and A_3.

For an arbitrary number k of mutually independent events, the features found above with $k = 3$ may be found to hold generally (see Exercises 14 and 15). Mutual independence of $A_1, A_2, \ldots, A_k$ will again, as with $k = 2$, imply the mutual independence of $A_1', A_2, \ldots, A_k$. What this comes down to is the verification of $P(A_1' \cap A_{i_1} \cap \ldots \cap A_{i_r}) = P(A_1') \prod_{j=1}^{r} P(A_{i_j})$ for any selection $A_{i_1}, A_{i_2}, \ldots, A_{i_r}$ of one or more events from $A_1, A_2, \ldots, A_k$. We may write

$$\begin{aligned} P(A_1' \cap A_{i_1} \cap A_{i_2} \cap \cdots \cap A_{i_r}) &= P(A_{i_1} \cap A_{i_2} \cap \cdots \cap A_{i_r}) - P(A_1 \cap A_{i_1} \cap A_{i_2} \cap \cdots \cap A_{i_r}) \\ &= \prod_{j=1}^{r} P(A_{i_j}) - P(A_1) \prod_{j=1}^{r} P(A_{i_j}) \end{aligned}$$

by making use of the mutual independence of $A_1, A_2, \ldots, A_k$ and the conclusion follows as before. The full import, which may be obtained by induction, is that the mutual independence of $A_1, A_2, \ldots, A_k$ implies the mutual independence of any k events chosen successively to be A_1 or A_1', A_2 or A_2', $\ldots$, A_k or A_k'. There are, as we know, 2^k such choices.

Example 4.5

Imagine again a sequence of seven flips of a coin. Let Ω be the set of all sequences of heads and tails of length seven, and suppose P is equally likely on Ω. Suppose we let $A_i = \{\omega | \ \omega \text{ has heads in the } i\text{th position}\}$ for

$i = 1, 2, \ldots, 7$. In this setup, the events $A_1, A_2, \ldots, A_7$ are mutually independent events. To see that this is so, we must demonstrate that for any selection of two or more events, say $A_{i_1}, A_{i_2}, \ldots, A_{i_r}$, $P(\bigcap_{j=1}^{r} A_{i_j}) = \prod_{j=1}^{r} P(A_{i_j})$. Now $P(A_i)$ is readily found to be $\frac{1}{2}$ for there are 2^6 sequences with a head in the ith position. Thus, $\prod_{j=1}^{r} P(A_{i_j}) = (\frac{1}{2})^r$. On the other hand, $\bigcap_{j=1}^{r} A_{i_j}$ is the set of all sequences that have heads in the i_1th position, the i_2th position, $\ldots$, the i_rth position. There are as many of these sequences as there are ways to fill up the remaining $7 - r$ positions, viz., 2^{7-r} ways. Hence,

$$P(\bigcap_{j=1}^{r} A_{i_j}) = \frac{2^{7-r}}{2^7} = 2^{-r} = \left(\frac{1}{2}\right)^r$$

and $A_1, A_2, \ldots, A_7$ are mutually independent. The events A_i', $i = 1, 2, \ldots, 7$ (the sequence has a tail in the ith position) are also mutually independent as are, for example, $A_1, A_2', A_3, A_4', A_5, A_6', A_7$.

There is no particular reliance on the number of tosses, 7, in this example. The mutual independence found is a general phenomenon for this type of situation and will be exploited more fully in Chapter Six.

EXERCISES 4.4

1. Suppose P is the equally likely measure on the events of $\Omega = \{\omega_1, \omega_2, \omega_3, \omega_4\}$. Identify three pairs of events A and B that are independent (A and B both proper subsets of Ω).

2. Suppose P is a probability measure on the events of $\Omega = \{\omega_1, \omega_2, \omega_3, \omega_4\}$ determined by $P(\{\omega_1\}) = \frac{1}{6}$, $P(\{\omega_2\}) = \frac{1}{12}$, $P(\{\omega_3\}) = \frac{1}{2}$, $P(\{\omega_4\}) = \frac{1}{4}$. Identify three pairs of events A and B that are independent (A and B both proper subsets of Ω).

3. Suppose P is the equally likely probability measure on the events of $\Omega = \{\omega_1, \omega_2, \omega_3, \omega_4, \omega_5\}$. Show that two events A and B from Ω are dependent unless at least one of them is $\varnothing$ or Ω. Generalize.

4. Suppose A and B are independent events with $P(A) = \frac{1}{6}$, $P(B) = \frac{1}{4}$. Determine $P(A \cup B)$, $P(A' \cap B)$, and $P(A' \cup B')$.

5. Suppose A and B are independent events and $P(A) = \frac{1}{6}$, $P(B) = \frac{1}{4}$. Determine the probability that precisely one of A and B occurs, that at most one of A and B occurs, that neither of A and B occurs.

6. Let P be equally likely on the sample space Ω of descriptions of all five-card hands from a deck of 52. Which of the following pairs of events do you think are independent?

(a) Having the ace of spades and having exactly four black cards.
(b) Having the ace of spades and having more black cards than red.
(c) Having exactly one 6 and having more black cards than red.
(d) Having exactly two spades and exactly two hearts and having more black cards than red.
(e) Having at most one card in each suit and having the ace of spades.
(f) Having at least four spades and having five spades.
(g) Having no more than one spade and having the ace of spades.

Show your independence guesses to be correct.

7. What may be said of an event A that is independent of itself?

8. Suppose a die is thrown three times in succession and that all 6^3 sequences of faces are equally likely. Which of the following pairs of events are independent?
(a) Having a first toss of two and having a second toss of three.
(b) Having a first toss of two and having the sum of the second and third tosses be six.
(c) Having a first toss of two and having at least two tosses of two.
(d) Having a first toss of two and having all three tosses be two.
(e) Having a first toss of two and having three tosses the same.
(f) Having a first toss of two and having the sum of three tosses be four.

9. Suppose $A_1, A_2, \ldots, A_k$ are mutually independent events each with probability $\frac{1}{3}$. What is the probability that none of these events occur? That exactly one of them occurs?

10. Events $A_1, A_2, \ldots, A_k$ are called *pairwise independent* if A_i and A_j are independent for every $i \neq j$. Show that if $A_1, A_2, \ldots, A_k$ are mutually independent then they are pairwise independent.

11. Let Ω consist of all 26^3 three-letter words, and let P be the equally likely probability measure on the events of Ω. If $A = \{\omega | \omega \text{ begins with s}\}$, $B = \{\omega | \omega \text{ has an s in the middle entry}\}$, and $C = \{\omega | \omega \text{ has two letters the same, one different}\}$, show that A, B, and C are pairwise independent but not mutually independent.

12. Show that if A and B_1 are independent, A and B_2 are independent, and $B_1 \cap B_2 = \emptyset$, then A and $B_1 \cup B_2$ are independent. Generalize.

13. Find a simple counter example to the statement of Exercise 12 if the disjointness of B_1 and B_2 is left out.

14. Show that the mutual independence of $A_1, A_2, \ldots, A_k$ implies the mutual independence of $A_2, A_3, \ldots, A_k$. Generalize.

15. Show that the mutual independence of $A_1, A_2, \ldots, A_k$ implies the mutual independence of $A_1 \cap A_2, A_3, A_4, \ldots, A_k$. Generalize.

16. Show that when mutual independence is replaced by pairwise independence Exercise 14 remains true while Exercise 15 does not.

17. Show that if $B_1, A_1, A_2, \ldots, A_k$ are mutually independent, $B_2, A_1, A_2, \ldots, A_k$ are mutually independent, and $B_1 \cap B_2 = \varnothing$, then $B_1 \cup B_2, A_1, A_2, \ldots, A_k$ are mutually independent. Generalize.

4.5 Partitions and independence

The first topic of this section is a set-theoretic one, viz., partitions. At this point, no benefit will accrue from discussing it in the language of sets as opposed to the probability language that we have been using. Thus, we will make the identification of a universal set with a sample space and of a subset with an event. A fixed sample space $\Omega = \{\omega_1, \omega_2, \ldots, \omega_n\}$ will be assumed throughout, and no probability measure will be imposed on it for the present discussion. The interaction of partitions with a probability measure is treated in the latter portions of the section.

Definition 4.4 A set of events $\{A_1, A_2, \ldots, A_k\}$ from Ω is called a *partition* of Ω if $A_1, A_2, \ldots, A_k$ are mutually exclusive and exhaustive events.

We shall talk of more than one partition of Ω, and we will designate them generally, by $\mathcal{A}$, $\mathcal{B}$, $\mathcal{C}$, etc. Typically, $\mathcal{A} = \{A_1, A_2, \ldots, A_n\}$, $\mathcal{B} = \{B_1, B_2, \ldots, B_m\}$, $\mathcal{C} = \{C_1, C_2, \ldots, C_q\}$, etc.

We are already aware of various partitions of Ω. $\{\Omega\}$ is one as is $\{A, A'\}$ for any event A from Ω. Another is the set of all the simple events $\{\{\omega_1\}, \{\omega_2\}, \ldots, \{\omega_n\}\}$. As soon as Ω is made more definite, others may be easily seen. For example, if Ω is the set of all descriptions of five-card hands from a deck of 52, then the events $A_i = \{\omega |\ \omega$ has in it i aces$\}$, $i = 0, 1, 2, 3, 4$ form a partition of Ω. So also do the events $B_i = \{\omega|\ \omega$ has in it i red cards$\}$, $i = 0, 1, 2, 3, 4, 5$, and $C_i = \{\omega|\ \omega$ has in it i spades$\}$, $i = 0, 1, 2, 3, 4, 5$.

The basic property of a partition $\mathcal{A} = \{A_1, A_2, \ldots, A_k\}$ of Ω is that each sample point belongs to one and only one of the member events of $\mathcal{A}$. In this sense, we can imagine each sample point being inspected and classified into one of the events A_i, $i = 1, 2, \ldots, k$. Thus, a five-card hand may be classified as to the number of aces it contains, the number of red cards it contains, or the number of spades it contains. It could also be classified as to the number of aces *and* the number of red cards it contains. A partition that would do this might consist of the events $A_i \cap B_j$ for $i = 0, 1, 2, 3, 4$ and $j = 0, 1, 2, 3, 4, 5$. For example, $A_0 \cap B_3$ is the set of hands having no aces and three red cards. We have been slightly generous in that some of the events $A_i \cap B_j$ are empty. For instance, it is not possible to have three aces and no red cards. Nonetheless, the events $A_i \cap B_j$, $i = 0, 1, 2, 3, 4$, and $j = 0, 1, 2, 3, 4, 5$ are mutually exclusive and exhaustive, and the set of them forms a partition of Ω.

When $\mathcal{A}_1, \mathcal{A}_2, \ldots, \mathcal{A}_r$ are all partitions of a sample space Ω, this same device may be used to form another partition of Ω.

Definition 4.5 The set of all events of the form $A_1 \cap A_2 \cap \cdots \cap A_r$ where $A_i \in \mathcal{A}_i$ for $i = 1, 2, \ldots, r$ is called the *product partition* of Ω induced by $\mathcal{A}_1, \mathcal{A}_2, \ldots, \mathcal{A}_r$.

We shall write the product partition induced by $\mathcal{A}_1, \mathcal{A}_2, \ldots, \mathcal{A}_r$ as $\mathcal{A}_1 \wedge \mathcal{A}_2 \wedge \cdots \wedge \mathcal{A}_r$ or as $\bigwedge_{i=1}^{r} \mathcal{A}_i$. Notice that it does not depend on the order in which the $\mathcal{A}_i$'s are taken and that, as above, some of the events $A_1 \cap A_2 \cap \cdots \cap A_r$ may be empty.

Example 4.6

Let Ω be the set of all people residing in the United States. Typically, a partition of Ω will arise whenever we imagine one or more characteristics of people that will break the population up into disjoint groups. Sex is such a characteristic, and $\{M, M'\}$ is a partition of Ω where $M = \{\omega | \ \omega \text{ is a male}\}$. Age in years is another, and $\{A_0, A_1, A_2, \ldots, A_{1000}\}$ is a partition of Ω where $A_i = \{\omega | \ \omega \text{ is } i \text{ years old}\}$. The product partition of these two is $\{M \cap A_0, M \cap A_1, \ldots, M \cap A_{1000}, M' \cap A_0, M' \cap A_1, \ldots, M' \cap A_{1000}\}$ wherein both characteristics are being accounted for and where $M' \cap A_{53}$ is the subset of 53-year-old females, for example. The business of filling out a questionnaire for employment, income tax, the Census Bureau, or the like might be viewed as the process of classifying oneself with respect to some partition of Ω. The information given may classify one with respect to sex, age, education, income, marriage status, geographical location, and color of eyes to name but a few possibilities.

Now let P be a probability measure on the events of Ω. A second form of independence is to be introduced, the independence of partitions. It is based directly on the independence of events as defined in Section 4.4. The point of difference that exists between the two forms may be described by saying that independence of events is independence within some set of events while independence of partitions is independence between sets of events.

Suppose $\mathcal{A} = \{A_1, A_2, \ldots, A_k\}$ and $\mathcal{B} = \{B_1, B_2, \ldots, B_m\}$ are two partitions of Ω.

Definition 4.6 $\mathcal{A}$ and $\mathcal{B}$ are called *independent partitions* if for every $i = 1, 2, \ldots, k$ and $j = 1, 2, \ldots, m$, the events A_i and B_j are independent.

To check the independence of $\mathcal{A}$ and $\mathcal{B}$, we must verify the $k \cdot m$ equations $P(A_i \cap B_j) = P(A_i)P(B_j)$, $i = 1, 2, \ldots, k$, $j = 1, 2, \ldots, m$. The number of these conditions may make independence of partition appear to be a quite restrictive property—in fact, it is. However, we shall meet many instances of independent partitions. For a simple example, imagine A and B to be any independent events from Ω. It is easy to see that $\{A, A'\}$ and $\{B, B'\}$ are independent partitions of Ω.

For the mutual independence of the partitions $\mathcal{A}_1, \mathcal{A}_2, \ldots, \mathcal{A}_k$ of Ω, we employ the concept of mutual independence of events.

Definition 4.7 $\mathcal{A}_1, \mathcal{A}_2, \ldots, \mathcal{A}_k$ are called *mutually independent partitions* if for every selection of an event A_1 from $\mathcal{A}_1$, an event A_2 from $\mathcal{A}_2, \ldots$, an event A_k from $\mathcal{A}_k$, the events $A_1, A_2, \ldots, A_k$ are mutually independent.

This concept of mutual independence reduces to what has been called independence when $k = 2$. An example of mutually independent partitions can be constructed in the same way as was done above for $k = 2$. If $A_1, A_2, \ldots, A_k$ are mutually independent events from Ω, then $\{A_1, A_1'\}$, $\{A_2, A_2'\}, \ldots, \{A_k, A_k'\}$ are mutually independent partitions. This is the case with the coin tossing example of Section 4.4, and the mutually independent partitions are those which divide sequences of heads and tails according to whether they are heads or tails in a given position.

EXERCISES 4.5

1. Let Ω be the set of all 26^3 three-letter words. Which of the following sets of events partition Ω?
   (a) $\mathcal{A} = \{A_1, A_2\}$ with $A_1 = \{\omega | \omega$ has more vowels than consonants$\}$, $A_2 = \{\omega | \omega$ has more consonants than vowels$\}$.
   (b) $\mathcal{A} = \{A_1, A_2, A_3\}$ with $A_i = \{\omega | \omega$ has i vowels$\}$, $i = 1, 2, 3$.
   (c) $\mathcal{A} = \{A_1, A_2, A_3\}$ with $A_i = \{\omega | \omega$ has i vowels$\}$, $i = 1, 2$, and $A_3 = \{\omega | \omega$ is all vowels or all consonants$\}$.
   (d) $\mathcal{A} = \{A_1, A_2, A_3\}$ with $A_i = \{\omega | \omega$ has $i - 1$ vowels$\}$, $i = 1, 2$, $A_3 = \{\omega | \omega$ has more vowels than consonants$\}$.

2. Let P be the equally likely probability measure on the events of $\Omega = \{\omega_1, \omega_2, \omega_3, \omega_4, \omega_5, \omega_6\}$. Find two pairs of partitions of Ω that are independent.

3. Let Ω be the set of all sequences of 1, 2, 3, 4, 5, 6 of length three corresponding to three consecutive tosses of a die, and let P be equally likely on the events of Ω. Let $\mathcal{A} = \{A_1, A_2, \ldots, A_6\}$, $\mathcal{B} = \{B_1, B_2, \ldots, B_6\}$ and $\mathcal{C} = \{C_1, C_2, \ldots, C_6\}$ where $A_i = \{\omega | \ \omega \text{ begins with } i\}$, $B_i = \{\omega | \ \omega \text{ has an } i \text{ in the middle position}\}$, and $C_i = \{\omega | \ \omega \text{ ends with } i\}$, $i = 1, 2, 3, 4, 5, 6$. Show that $\mathcal{A}$, $\mathcal{B}$, and $\mathcal{C}$ are mutually independent partitions of Ω.

4. Show that if $\mathcal{A}$ is any partition of a sample space Ω on which P is a probability measure, then $\{\Omega\}$ and $\mathcal{A}$ are independent partitions of Ω.

5. Show that if $\mathcal{A}$ is a partition of a sample space $\Omega = \{\omega_1, \omega_2, \ldots, \omega_n\}$ on which P is a probability measure for which the partitions $\mathcal{A}$ and $\{\{\omega_1\}, \{\omega_2\}, \ldots, \{\omega_n\}\}$ are independent, then for every $A \in \mathcal{A}$, $P(A) = 1$ or $P(A) = 0$.

6. Show that $\mathcal{A} = \{A_1, A_2, \ldots, A_k\}$ is a partition of a sample space Ω if and only if $\bar{\mathcal{A}} = \{A_1, A_2, \ldots, A_k, \varnothing\}$ is a partition of Ω.

7. Define a concept of pairwise independence of partitions $\mathcal{A}_1, \mathcal{A}_2, \ldots, \mathcal{A}_k$ by making use of the notion of pairwise independence of events (see Exercise 10 of Section 4.4). Does it then follow that if $\mathcal{A}_1, \mathcal{A}_2, \ldots, \mathcal{A}_k$ are mutually independent partitions, they are pairwise independent? What can be said about the reverse implication (see Exercise 11 of Section 4.4)?

8. Show that if $\mathcal{A}_1, \mathcal{A}_2, \ldots, \mathcal{A}_k$ are mutually independent partitions, then so are $\mathcal{A}_2, \mathcal{A}_3, \ldots, \mathcal{A}_k$ (see Exercise 14 of Section 4.4). Generalize.

9. Show that if $\mathcal{A}_1, \mathcal{A}_2, \ldots, \mathcal{A}_k$ are mutually independent partitions, then so are $\mathcal{A}_1 \wedge \mathcal{A}_2, \mathcal{A}_3, \mathcal{A}_4, \ldots, \mathcal{A}_k$ (see Exercise 15 of Section 4.4). Generalize.

10. What may be said of a partition $\mathcal{A}$ that is independent of itself?

***11.** Let $\mathcal{A} = \{A_1, A_2, \ldots, A_k\}$ and $\mathcal{B} = \{B_1, B_2, \ldots, B_m\}$ be partitions of a sample space $\Omega = \{\omega_1, \omega_2, \ldots, \omega_k\}$. $\mathcal{A}$ is said to be *coarser* than $\mathcal{B}$ (or $\mathcal{B}$ is finer than $\mathcal{A}$) provided every B_j, $j = 1, 2, \ldots, m$ is contained in some one of $A_1, A_2, \ldots, A_k$.

(a) Show that if $\mathcal{A}$ is coarser than $\mathcal{B}$, then every $B_j \neq \varnothing$ is contained in precisely one of $A_1, A_2, \ldots, A_k$ and every $A_i \neq \varnothing$ is the union of those events B_j that it contains.

(b) Show that for any two partitions $\mathcal{A}$ and $\mathcal{B}$ of Ω, $\mathcal{A} \wedge \mathcal{B}$ is finer than $\mathcal{A}$.

(c) Show that any partition $\mathcal{A}$ of Ω is finer than $\{\Omega\}$ and coarser than $\{\{\omega_1\}, \{\omega_2\}, \ldots, \{\omega_n\}\}$.

***12.** Let Ω be a sample space, and let P be a probability measure on the events of Ω. Show that if $\mathcal{A}$ and $\mathcal{B}$ are independent partitions of Ω and $\mathcal{C}$ is a coarser partition of Ω than $\mathcal{B}$, then $\mathcal{A}$ and $\mathcal{C}$ are independent. (See Exercise 12 of Section 4.4.)

***13.** Let Ω be a sample space. An *algebra* of events of Ω is a nonempty set of events $\mathbf{A} = \{A_1, A_2, \ldots, A_k\}$ for which (i) $A'_i \in \mathbf{A}$ for $i = 1, 2, \ldots, k$; (ii) $A_i \cup A_j \in \mathbf{A}$ for $i = 1, 2, \ldots, k$ and $j = 1, 2, \ldots, k$. Show that if $\mathbf{A} = \{A_1, A_2, \ldots, A_k\}$ is an algebra of events from Ω, then

(a) $\Omega \in \mathbf{A}$.
(b) $A_i \cap A_j \in \mathbf{A}$ for $i = 1, 2, \ldots, k$ and $j = 1, 2, \ldots, k$.
(c) $\bigcup_{i=1}^{r} A_i \in \mathbf{A}$ for $1 \leq r \leq k$.

***14.** Show that if $\mathcal{A} = \{A_1, A_2, \ldots, A_k\}$ is a partition of Ω, then $\mathbf{A} = \{A \mid A = \varnothing$ or $A = \bigcup_{j=1}^{r} A_{i_j}$ for some selection $A_{i_1}, A_{i_2}, \ldots, A_{i_r}$ of one or more events from $\mathcal{A}\}$ is an algebra of events from Ω. Show that if $\mathbf{A}$ is an algebra of events from Ω, then $\mathbf{A}$ arises from some partition $\mathcal{A}$ in just this way.

***15.** Let $\mathbf{A}$ and $\mathbf{B}$ be two algebras of events of a sample space Ω on which P is a probability measure. $\mathbf{A}$ and $\mathbf{B}$ are called independent if for each $A \in \mathbf{A}$ and $B \in \mathbf{B}$, A and B are independent events. Let $\mathcal{A}$ and $\mathcal{B}$ be two independent partitions of Ω, and let $\mathbf{A}$ and $\mathbf{B}$ be algebras formed from $\mathcal{A}$ and $\mathcal{B}$ as in Exercise 14. Show that $\mathbf{A}$ and $\mathbf{B}$ are independent.

CHAPTER FIVE

Random Variables

5.1 Introduction

In this chapter we shall discuss the notion of a random variable. Roughly speaking, a random variable is a numerical quantity whose value depends on the outcome of some experiment. We shall see from examples below that situations to which probability theory is applicable almost always lead us to ask questions about the behavior of such quantities, which is why they are a natural part of our study.

We shall first define random variables in the mathematical framework and study the various operations to which random variables may be subjected. Later in this chapter, we shall study questions concerned with the probabilistic behavior of random variables.

5.2 Random variables

When an experiment is performed, we are often less interested in the specific outcome of it than in some aspect of the outcome that may be shared with other particular outcomes. Here, this aspect will take the form of a numerical quantity that depends on the outcome. We might think of such a quantity as a "measurement" on the outcome—it provides a partial description of an outcome in the same way that, say, age in years provides a partial description of a person. Let us consider some examples.

Example 5.1

Harry and Tom play a coin-tossing game. A coin is flipped 100 times, and, by agreement, each time a head appears Tom pays Harry $1 and each time a tail appears Harry pays Tom $1. A detailed sample space for this experiment might then be the Cartesian product of the set {H, T} with itself 100 times. A sample point here is a 100-tuple of H's and T's, and there are 2^{100} sample points in all.

From such a detailed description of the outcome, we may draw a wealth of information. For example, from a given sample point we may determine the outcome of the thirty-seventh toss, the total number of heads on even numbered tosses, and the leading money winner after 65 tosses. As far as the game is concerned, however, Harry may even regard these pieces of information as secondary in importance to the amount of money he wins or loses. He realizes that if h heads appear in all, then there are $100 - h$ tails, and his gain (or loss) in dollars will be $h - (100 - h) = 2h - 100$. This numerical quantity depends on the outcome in a definite way, and it ranges between -100 ($h = 0$) and $+100$ ($h = 100$) in increments of 2.

Example 5.2

Imagine the experiment of drawing a 13-card hand from a standard deck. A detailed sample space might consist of all possible complete descriptions of hands and thus contain $\binom{52}{13}$ sample points. From a given sample point, we could determine such things as the number of aces in the hand, the number of spades in the hand, and whether or not the hand contains the six of clubs. A bridge player who is dealt such a hand views as important, among other things, the "worth" of the hand as measured by its honor count. The honor count of a hand is determined by counting 4 points for each ace, 3 points for each king, 2 points for each queen, and 1 point for each jack. Thus, if a hand contains a aces, k kings, q queens, and j jacks, the honor count is $4a + 3k + 2q + j$. This numerical quantity depends on the sample point (outcome) in a definite way, and it ranges between zero (the hand is a bust, i.e., contains no aces and no face cards) and 37 (the hand contains four aces, four kings, four queens, and a jack).

Example 5.3

In order to determine the efficacy of a newly developed serum against a type of tumor, a medical research worker injects the serum into 100 guinea pigs having the tumor. Suppose that each guinea pig is subsequently classified as either cured or not cured by the serum. A sample space might then be the Cartesian product of the set $\{c, n\}$ with itself 100 times. Sample points are then 100-tuples of c's and n's with a c (or n) in the ith position denoting, say, a cure (or not a cure) of the ith guinea pig. Such a sample point would indicate precisely which guinea pigs were cured and which were not. What is usually more important than knowing *which* were cured is knowing *how many* were cured. Such a numerical quantity depends on the outcome in a definite way and ranges between zero [the outcome is $(n, n, \ldots, n)$] and 100 [the outcome is $(c, c, \ldots, c)$].

In these examples, our interest has been focused on some quantity taking values dependent upon the outcome of an experiment. Given the outcome (sample point), there was in each case a definite rule that enabled us to decide what the value of the quantity was. In statistical usage, such a quantity is called a *random variable,* doubtless because its value is capable of varying from one outcome to the next (thus, the word "variable") and the outcome itself depends on chance (thus, the word "random"). The mathematical concept that describes these quantities is that of a *real-valued function* defined on a sample space. We therefore make the following definition.

Definition 5.1 A *random variable defined on a sample space* Ω is a real-valued function defined on Ω.

The name "random variable" for a function defined on Ω is generally acknowledged to be a poor choice. It has, however, the sanction of common usage. For a fixed sample space Ω, there are many random variables defined on Ω. Generally, we shall designate random variables by the letters W, X, Y, and Z.

A particular random variable X on Ω may be completely described by making a table of its values like the one shown in Figure 5.1. However, it is clear that if Ω has a large number of sample points, this is not a practical description. In such cases, we may specify X by describing the rule whereby $X(\omega)$ may be determined for each sample point ω. For example, Harry's dollar gain X in Example 5.1 may be described by saying "X is the number of heads minus the number of tails in ω," or "X is

Sample points	ω_1	ω_2	...	ω_n
Values of X	$X(\omega_1)$	$X(\omega_2)$	...	$X(\omega_n)$

FIGURE 5.1

the number of heads minus the number of tails in 100 coin tosses." How would you similarly describe other random variables in the settings of Examples 5.1, 5.2, and 5.3?

Given a random variable X on Ω, we will often specify other random variables by making use of the specification of X itself. For example, let us take Harry's dollar gain X as above. By the random variable $X/2 + 50$ we shall mean the random variable that at any point ω has the value $X(\omega)/2 + 50$. In the context of the coin-tossing game, $X/2 + 50$ can then be interpreted as the number of heads that appear in the sequence of 100 tosses. In the same vein, we might speak of $X + 150$ and $2X + 150$ as random variables on Ω. These are random variables, respectively, with values $X(\omega) + 150$ and $2 \cdot X(\omega) + 150$ at a given point ω. $X + 150$ would admit an interpretation as Harry's bankroll in dollars after the game if he enters the game with \$150. $2X + 150$ would be his bankroll in dollars after the game if he begins with \$150 and the stakes are doubled. We may wish to consider more random variables of this type than the number of interpretations we can think up relative to the underlying experiment. For example, we may wish to consider the random variables X^2 and $X^3 + 5$. Having described X as Harry's dollar gain, we could describe these as the square of Harry's dollar gain and the cube of Harry's dollar gain plus 5. Formally, these are random variables on Ω with values $[X(\omega)]^2$ and $[X(\omega)]^3 + 5$ at any point ω, respectively.

In general, if X is a random variable on a sample space Ω and if g is a real-valued function of a real variable, we define a random variable $g(X)$ on Ω by, for each $\omega \in \Omega$,

$$(g(X))(\omega) = g(X(\omega)) \tag{5.1}$$

All the examples above are simply special cases of Eq. (5.1) for appropriate choices of functions g. In order of their appearance, the choices of g were those defined by, for each real number x, $g(x) = x/2 + 50$, $g(x) = x + 150$, $g(x) = 2x + 150$, $g(x) = x^2$, and $g(x) = x^3 + 5$. *A random variable defined by Eq.* (5.1) *is said to be a function of X.*

Given two or more random variables, say $X_1, X_2, \ldots, X_r$, on Ω, we can in the same way specify other random variables on Ω. In the context of Example 5.2, let us take X_1 to be the number of aces in the hand, X_2 to be the number of kings, X_3 to be the number of queens, and X_4 to be the number of jacks. By the random variable $4X_1 + 3X_2 + 2X_3 + X_4$ we shall mean the random variable on Ω that has the value $4X_1(\omega) + 3X_2(\omega) + 2X_3(\omega) + X_4(\omega)$ at each ω. This, as we have seen, is the honor count of the hand. Similarly, $X_1 + X_2 + X_3 + X_4$ is the random variable with value $X_1(\omega) + X_2(\omega) + X_3(\omega) + X_4(\omega)$ at each $\omega \in \Omega$. It may be interpreted as the total number of aces and face cards in the hand. We might also consider the random variables $X_1X_2X_3X_4$ and $X_1^2 + X_2^2 + X_3^2 + 5$. These take values $X_1(\omega)X_2(\omega)X_3(\omega)X_4(\omega)$ and $[X_1(\omega)]^2 + [X_2(\omega)]^2 + [X_3(\omega)]^2 + 5$ at each $\omega \in \Omega$, respectively.

In general, if $X_1, X_2, \ldots, X_r$ are all random variables on a sample space Ω and if g is a real-valued function of r real variables, we define the random variable $g(X_1, X_2, \ldots, X_r)$ on Ω by, for each $\omega \in \Omega$,

$$(g(X_1, X_2, \ldots, X_r))(\omega) = g(X_1(\omega), X_2(\omega), \ldots, X_r(\omega)) \tag{5.2}$$

The examples above are special cases of Eq. (5.2) for appropriate choices of functions g.

In order of their appearance, the choices of g were those defined by, for any real numbers x_1, x_2, x_3, and x_4,

$$g(x_1, x_2, x_3, x_4) = 4x_1 + 3x_2 + 2x_3 + x_4$$

$$g(x_1, x_2, x_3, x_4) = x_1 + x_2 + x_3 + x_4$$

$$g(x_1, x_2, x_3, x_4) = x_1x_2x_3x_4$$

and

$$g(x_1, x_2, x_3, x_4) = x_1^2 + x_2^2 + x_3^2 + 5$$

A random variable defined by Eq. (5.2) *is said to be a function of* $X_1, X_2, \ldots, X_r$.

We conclude this section by pointing out an important class of random variables on a sample space Ω. In particular, if A is any event of Ω, there is a random variable X defined on Ω that takes the value 1 at each sample point in A and takes the value 0 at each sample point in A'. Thus, the value of X at a given sample point ω indicates whether or not $\omega \in A$. For this reason, X is called the *indicator function* (or *indicator random variable*) of A; it will be denoted by I_A. Therefore,

$$\begin{aligned} I_A(\omega) &= 1 \quad \text{if } \omega \in A \\ &= 0 \quad \text{if } \omega \in A' \end{aligned} \tag{5.3}$$

Clearly, each event A of Ω gives rise to an indicator random variable all its own, so that there are 2^n of them when Ω has n sample points. Especially, it may be seen that I_Ω is 1 at every sample point, while $I_\varnothing$ is 0 at every sample point.

Indicator random variables may be thought of as encodings of information into numbers. In Example 5.2, for instance, we noted that it was possible to determine from a given sample point whether or not the hand contained the six of clubs. If this is a question of interest, the answer to it may be given in terms of the value of a random variable. We simply equate the occurrence of the event with the number 1 and its nonoccurrence with the number 0. Now, the question of interest is phrased in terms of the value of the indicator random variable I_A of the event $A = \{\omega | \omega$ contains the six of clubs$\}$. If its value is 1, the event A occurred; if its value is 0, the event A did not occur.

EXERCISES 5.2

1. Imagine the experiment of tossing a coin four times, and let the sample space Ω be the Cartesian product of the set $\{H, T\}$ with itself four times. Consider the random variables on Ω described by the following: X is the number of heads in ω, Y is the number of tails in ω, Z is the length of the longest block of consecutive tails in ω ($Z = 0$ if no tails are tossed), and W is the number of the toss in which the last tail is tossed ($W = 0$ if no tails are tossed).

(a) For each $\omega \in \Omega$, give the value of each of the random variables X, Y, Z, and W.

(b) For each $\omega \in \Omega$, give the value of each of the random variables $X + Y$, $X - Y$, $Z + W$, W^2, and XZ.

(c) Let g and h be functions of two variables defined by, for any real numbers x and y, $g(x, y) = |x - y|$ and $h(x, y) = \min (x, y)$ (the minimum of the two numbers x and y). For each $\omega \in \Omega$, give the value of each of the random variables $g(X, Z)$ and $h(X, Y)$.

(d) Is Y a function of X? Is X a function of Y? Is X a function of the two random variables $X - Y$ and $X + Y$?

2. Suppose X is a random variable on Ω that takes only the values 0 and 1. Show that X is the indicator random variable of some event A of Ω.

3. Let A and B be two events of Ω.

(a) Show that $I_A + I_{A'} = I_\Omega$, $I_{A \cap B} = I_A I_B$, and $I_{A \cup B} = I_A + I_B - I_{A \cup B}$.

(b) Give $I_{A \cap B'}$, $I_{A' \cap B}$, $I_{A' \cap B'}$, and $I_{A' \cup B'}$ in terms of I_A and I_B.

(c) If C is another event of Ω, give $I_{A \cup B \cup C}$ in terms of I_A, I_B, and I_C.

4. If X and Y are two random variables on a sample space Ω, show that $(X + Y)^2 = X^2 + 2XY + Y^2$ and $(X - Y)^2 = X^2 - 2XY + Y^2$.

5. Imagine the experiment of rolling a die twice, and let the sample space Ω be the Cartesian product of the set $\{1, 2, 3, 4, 5, 6\}$ with itself. Consider the random variables on Ω described as follows: X is the total score on the two rolls, and Y is the largest score on the two rolls.

(a) Utilizing the nature of the sample points, give rules whereby $X(\omega)$ and $Y(\omega)$ may be found at each $\omega \in \Omega$.

(b) Is there any evident relationship between X and Y?

(c) What are the possible values of X? of Y?

(d) Describe a random variable on Ω that has 36 possible values.

6. Suppose the die of Exercise 5 is rolled n times, and let Ω be the Cartesian product of the set $\{1, 2, 3, 4, 5, 6\}$ with itself n times. Answer parts **(a)**, **(b)**, and **(c)** of Exercise 5 if now X is the total score on the n rolls and Y is the largest score on the n rolls.

7. An urn contains 20 balls of which eight are black and twelve are red. Five balls are drawn in succession, the color of each is noted, and they are then discarded. Describe a sample space for this experiment as a set of five-tuples with appropriate entries. Let X be the number of black balls drawn.
(**a**) Give a rule whereby $X(\omega)$ may be determined for each sample point.
(**b**) What are the possible values of X?

8. An urn contains N balls of which b are black and r are red $(b + r = N)$. n balls are drawn in succession, the color of each is noted, and they are then discarded. Describe a sample space for this experiment as a set of n-tuples with appropriate entries. Let X be the number of black balls drawn, and answer parts (**a**) and (**b**) of Exercise 7.

5.3 Partitions determined by random variables

Let X be a random variable on the sample space $\Omega = \{\omega_1, \omega_2, \ldots, \omega_n\}$. Accordingly, X takes on the values $X(\omega_1), X(\omega_2), \ldots, X(\omega_n)$. If these numbers are all distinct, then X can take on n values in all. If they are not all distinct, X can take on less than n values in all. We suppose generally that X takes on the distinct values $x_1, x_2, \ldots, x_k$, $k \leq n$. Corresponding to each value x_i, $1 \leq i \leq k$, we define an event A_i of Ω by

$$A_i = \{\omega | X(\omega) = x_i\} \tag{5.4}$$

Thus, A_i is simply the event that X takes the value x_i.

It is easy to check that $\{A_1, A_2, \ldots, A_k\}$, each event defined as above, is a partition of Ω. For the first part, if these events were not mutually exclusive, we could find a sample point ω in both A_i and A_j for some $i \neq j$. But then we would have both $X(\omega) = x_i$ and $X(\omega) = x_j \neq x_i$. Secondly, since X is defined on Ω, each $\omega \in \Omega$ has the property that $X(\omega) = x_i$ for some one of $i = 1, 2, \ldots, k$; hence, $\bigcup_{i=1}^{k} A_i = \Omega$.

Definition 5.2 Let X be a random variable on a sample space Ω taking the distinct values $x_1, x_2, \ldots, x_k$, and let $A_i = \{\omega | X(\omega) = x_i\}$ for $i = 1, 2, \ldots, k$. The partition $\{A_1, A_2, \ldots, A_k\}$ of Ω is called *the partition of Ω induced by X* and is denoted by $\mathcal{A}_X$.

The knowledge of $\mathcal{A}_X$ together with the value that X takes at each sample point of a given event in the partition enables us to specify X in a manner superficially different from what we did in the last section. Thus, if X takes the values $x_1, x_2, \ldots, x_k$ and $\mathcal{A}_X = \{A_1, A_2, \ldots, A_k\}$

with $A_i = \{\omega | X(\omega) = x_i\}$ for $i = 1, 2, \ldots, k$, then

$$\begin{aligned} X(\omega) &= x_1 \quad \text{if } \omega \in A_1 \\ &= x_2 \quad \text{if } \omega \in A_2 \\ &\ \ \vdots \\ &= x_k \quad \text{if } \omega \in A_k \end{aligned} \tag{5.5}$$

Example 5.4

Let X be the total number of heads in four tosses of a coin on a sample space Ω that is the Cartesian product of the set {H, T} with itself four times. Written out in its entirety, the function X is as shown in Figure 5.2.

The distinct values taken by X are 0, 1, 2, 3, and 4. For these values, the events A_i are, respectively, $A_1 = \{(T, T, T, T)\}$, $A_2 = \{(H, T, T, T), (T, H, T, T), (T, T, H, T), (T, T, T, H)\}$, $A_3 = \{(H, H, T, T), (H, T, H, T), (H, T, T, H), (T, H, H, T), (T, H, T, H), (T, T, H, H)\}$, $A_4 = \{(H, H, H, T), (H, H, T, H), (H, T, H, H), (T, H, H, H)\}$, and $A_5 = \{(H, H, H, H)\}$. Therefore, $\mathcal{A}_X = \{A_1, A_2, A_3, A_4, A_5\}$. This partition is easily seen to be the partition induced by random variables other than X also. All we need do is, with some care, change the values on at least one of the A_i, $i = 1, 2, 3, 4, 5$. Consider, for example, the random variable $2X + 3$. This random variable takes the value 3 at each sample point of A_1, the value 5 at each sample point of A_2, the value 7 at each sample point of A_3, the value 9 at each sample point of A_4, and the value 11 at each sample point of A_5. In other words, $\mathcal{A}_{2X+3} = \mathcal{A}_X$. On the other hand, the random variable $|X - 2|$ takes the value 2 at each sample point of $A_1 \cup A_5$, the value 1 at each sample point of $A_2 \cup A_4$,

ω	$X(\omega)$	ω	$X(\omega)$
(H, H, H, H)	4	(T, H, H, H)	3
(H, H, H, T)	3	(T, H, H, T)	2
(H, H, T, H)	3	(T, H, T, H)	2
(H, T, H, H)	3	(T, T, H, H)	2
(H, H, T, T)	2	(T, H, T, T)	1
(H, T, H, T)	2	(T, T, H, T)	1
(H, T, T, H)	2	(T, T, T, H)	1
(H, T, T, T)	1	(T, T, T, T)	0

FIGURE 5.2

and the value 0 at each sample point of A_3. Thus,

$$\mathcal{A}_{|X-2|} = \{A_1 \cup A_5, A_2 \cup A_4, A_3\}$$

Consider now two random variables X and Y on a sample space Ω. Let us suppose that X takes the values $x_1, x_2, \ldots, x_k$ on the events $A_1, A_2, \ldots, A_k$, respectively, and that Y takes the values $y_1, y_2, \ldots, y_m$ on the events $B_1, B_2, \ldots, B_m$, respectively. We have seen that $\mathcal{A}_X = \{A_1, A_2, \ldots, A_k\}$ and $\mathcal{A}_Y = \{B_1, B_2, \ldots, B_m\}$ are both partitions of Ω. The behavior of X at a sample point is therefore one of the following: $X(\omega) = x_i$, $i = 1, 2, \ldots, k$. In particular, $X(\omega) = x_i$ if and only if $\omega \in A_i$. The same comments may be made about Y. The *joint* behavior of X and Y at a sample point is one of the following: $X(\omega) = x_i$ and $Y(\omega) = y_j$, $i = 1, 2, \ldots, k$, $j = 1, 2, \ldots, m$. In particular, $X(\omega) = x_i$ and $Y(\omega) = y_j$ if and only if $\omega \in A_i \cap B_j$. In discussing the joint behavior of X and Y we are thus led to the contemplation of the events $A_i \cap B_j$, $i = 1, 2, \ldots, k$, $j = 1, 2, \ldots, k$. The set of these, viz., $\{A_1 \cap B_1, A_1 \cap B_2, \ldots, A_1 \cap B_m, A_2 \cap B_1, \ldots, A_k \cap B_m\}$, is another partition of Ω, and, in fact, it is the product partition $\mathcal{A}_X \wedge \mathcal{A}_Y$. This product partition may be exhibited in the form of a double-entry table as in Figure 5.3.

In Figure 5.3, the entry in the ith row and jth column ($1 \leq i \leq k$, $1 \leq j \leq m$) is the event $A_i \cap B_j$, i.e., the event that $X = x_i$ and $Y = y_j$. The events in the ith row are $A_i \cap B_1, A_i \cap B_2, \ldots, A_i \cap B_m$ on each of which $X = x_i$. Moreover, the union of these events is just the event that $X = x_i$, for,

$$\begin{aligned}(A_i \cap B_1) \cup (A_i \cap B_2) \cup \cdots \cup (A_i \cap B_m)& \\ &= A_i \cap (B_1 \cup B_2 \cup \cdots \cup B_m) \\ &= A_i \cap \Omega = A_i \qquad (5.6)\end{aligned}$$

$X \backslash Y$	y_1	y_2	$\ldots$	y_m	$\mathcal{A}_X$
x_1	$A_1 \cap B_1$	$A_1 \cap B_2$	$\ldots$	$A_1 \cap B_m$	A_1
x_2	$A_2 \cap B_1$	$A_2 \cap B_2$	$\ldots$	$A_2 \cap B_m$	A_2
$\vdots$	$\vdots$	$\vdots$		$\vdots$	$\vdots$
x_k	$A_k \cap B_1$	$A_k \cap B_2$	$\ldots$	$A_k \cap B_m$	A_k
$\mathcal{A}_Y$	B_1	B_2	$\ldots$	B_m	Ω

FIGURE 5.3

In the same way, the events in the jth column are $A_1 \cap B_j, A_2 \cap B_j, \ldots, A_k \cap B_j$ on each of which $Y = y_j$. The union of these events is the event that $Y = y_j$, i.e., $(A_1 \cap B_j) \cup (A_2 \cap B_j) \cup \cdots \cup (A_k \cap B_j) = B_j$. Thus, in Figure 5.3, the result of forming a union across a row is given in the last column, and the result of forming a union down a column is given in the last row. We also note that Ω is the union of the events in the last column, and, at the same time, it is the union of the events in the last row.

In an analogous fashion, the joint behavior of random variables $X_1, X_2, \ldots, X_r$ on a sample space Ω leads to the events of the product partition $\mathcal{A}_{X_1} \wedge \mathcal{A}_{X_2} \wedge \cdots \wedge \mathcal{A}_{X_r}$. On each event from this partition, the random variables $X_1, X_2, \ldots, X_k$ take some set of values, and these values differ between different events.

Example 5.5

Suppose X is the number of black aces and Y is the number of red cards in a hand of five drawn from an ordinary deck, both defined on a sample space Ω of descriptions of all possible $\binom{52}{5}$ hands. X takes values 0, 1, and 2 on events A_1, A_2, and A_3, respectively, while Y takes values 0, 1, 2, 3, 4, and 5 on events B_1, B_2, B_3, B_4, B_5, and B_6, respectively. The events A_i may be specified by simply writing $A_i = \{\omega | \omega \text{ contains } i - 1 \text{ black aces}\}$, $i = 1, 2, 3$. Similarly, the events B_j are specified by writing $B_j = \{\omega | \omega \text{ contains } j - 1 \text{ red cards}\}$, $j = 1, 2, \ldots, 6$.

On the event $A_i \cap B_j$, we have $X = i - 1$ and $Y = j - 1$, i.e., the hand contains $i - 1$ black aces and $j - 1$ red cards. Some of these events are empty inasmuch as the hand contains only five cards. In particular, $A_2 \cap B_6 = \varnothing$, $A_3 \cap B_6 = \varnothing$, and $A_3 \cap B_5 = \varnothing$. Each sample point must then fall into one of the remaining 15 events of $\mathcal{A}_X \wedge \mathcal{A}_Y$.

Now together with X and Y consider Z to be the number of red queens in the hand. Z takes values 0, 1, and 2 on the events C_1, C_2, and C_3, respectively, where $C_q = \{\omega | \omega \text{ contains } q - 1 \text{ red queens}\}$, $q = 1, 2, 3$. The product partition $\mathcal{A}_X \wedge \mathcal{A}_Y \wedge \mathcal{A}_Z$ now consists of $3 \cdot 6 \cdot 3 = 54$ events of the form $A_i \cap B_j \cap C_q$. On $A_i \cap B_j \cap C_q$, we have $X = i - 1$, $Y = j - 1$, and $Z = q - 1$. Many of these events are empty for two reasons, viz., the hand contains only five cards, hence $A_i \cap B_j \cap C_q = \varnothing$ if $i + j - 2 > 5$, and the hand cannot contain more red queens than red cards, hence $A_i \cap B_j \cap C_q = \varnothing$ if $q > j$.

EXERCISES 5.3

1. Let the sample space Ω for throwing a die twice be the Cartesian product of the set $\{1, 2, 3, 4, 5, 6\}$ with itself. Consider the random variables defined on Ω by the following: X is the score on the first throw, Y is the score on the second throw, and Z is the total score.

(a) Give the partitions $\mathcal{A}_X$, $\mathcal{A}_Y$, and $\mathcal{A}_Z$.
(b) Give the partition $\mathcal{A}_X \wedge \mathcal{A}_Z$ in the form of a double-entry table, and then verify Eq. (5.6) for this case.
*(c) Show that $\mathcal{A}_Z$ is coarser than $\mathcal{A}_X \wedge \mathcal{A}_Y$ (see Exercise 11 of Section 4.5).

2. In the setting of Exercise 1, let W be the random variable $X - Y$. Give the partitions $\mathcal{A}_W$ and $\mathcal{A}_{W^2}$. What relations can you detect between the events of $\mathcal{A}_W$ and those of $\mathcal{A}_{W^2}$?

3. Give a verbal explanation of Eq. (5.6) as it relates to the values of X and Y.

4. Let A and B be events of a sample space Ω.
(a) Give the partitions $\mathcal{A}_{I_A}$, $\mathcal{A}_{I_B}$, and $\mathcal{A}_{I_A} \wedge \mathcal{A}_{I_B}$.
(b) What is the relationship between $\mathcal{A}_{I_A}$ and $\mathcal{A}_{I_{A'}}$?
(c) Give the partitions $\mathcal{A}_{I_\varnothing}$ and $\mathcal{A}_{I_\Omega}$.

5. Let X be a random variable on Ω, and let $\mathcal{C} = \{C_1, C_2, \ldots, C_q\}$ be any partition of Ω. Suppose for each $\omega \in C_i$, $X(\omega) = z_i$, $i = 1, 2, \ldots, q$ (the numbers $z_1, z_2, \ldots, z_q$ need not be distinct).
(a) Show that $X = \sum_{i=1}^{q} z_i I_{C_i}$.
(b) Apply part (a) when $\mathcal{C}$ is the partition $\mathcal{A}_X$.

***6.** Generalize Eq. (5.6) to the case of r random variables.

***7.** Let X be a random variable on a sample space Ω, and let $g(X)$ be a function of X. Show that $\mathcal{A}_{g(X)}$ is coarser than $\mathcal{A}_X$. Conversely, let Y be a random variable on Ω for which $\mathcal{A}_Y$ is coarser than $\mathcal{A}_X$. Show that there is a function g for which $Y = g(X)$.

***8.** Let X, Y, and Z be random variables on a sample space Ω. Show that there is a function g for which $Z = g(X, Y)$ if and only if $\mathcal{A}_Z$ is coarser than $\mathcal{A}_X \wedge \mathcal{A}_Y$. Generalize this to a criterion for deciding if the random variable Z is a function of r given random variables $X_1, X_2, \ldots, X_r$.

***9.** State the results of Exercises 7 and 8 in terms of the algebras generated by the partitions in question.

***10.** Let X be a random variable on a sample space Ω that assumes the values $x_1, x_2, \ldots, x_k$. Show that $\mathcal{A}_{g(X)} = \mathcal{A}_X$ if and only if $g(x_i) = g(x_j)$ implies $x_i = x_j$. Use this result to get conditions on $x_1, x_2, \ldots, x_k$ so that $\mathcal{A}_{X^2} = \mathcal{A}_X$.

5.4 Probability distributions

Up to this point, random variables have been treated simply as functions on a sample space Ω. It is now time to take into account the fact that Ω may be equipped with a probability measure. Therefore, we now assume P is a probability measure on the events of Ω.

Consider first a single random variable X on Ω. Let $x_1, x_2, \ldots, x_k$ be the distinct values of X, and let $A_i = \{\omega | X(\omega) = x_i\}$, $i = 1, 2, \ldots, k$ be the events of $\mathfrak{A}_X$.

Definition 5.3 The function f_X defined on the set $\{x_1, x_2, \ldots, x_k\}$ by $f_X(x_i) = P(A_i)$, $i = 1, 2, \ldots, k$, is called the *probability distribution* of X.

To specify the probability distribution of X, we must specify the values $x_1, x_2, \ldots, x_k$ of X and the probabilities $P(A_1), P(A_2), \ldots, P(A_k)$. Of course, $P(A_i)$ is for each i just the probability of the event that $X = x_i$. Clearly, we have $f_X(x_i) \geq 0$ for each i and further

$$\sum_{i=1}^{k} f_X(x_i) = \sum_{i=1}^{k} P(A_i) = P\left(\bigcup_{i=1}^{k} A_i\right) = P(\Omega) = 1$$

Example 5.6

We return to Example 5.4 wherein we talked of X as the number of heads in four tosses of a coin. Let us suppose that P is the equally likely probability measure on the events of Ω as it is set out in Figure 5.2. We then find that

$$P(A_1) = P(\{\omega | X(\omega) = 0\}) = f_X(0) = \tfrac{1}{16}$$

$$P(A_2) = P(\{\omega | X(\omega) = 1\}) = f_X(1) = \tfrac{4}{16} = \tfrac{1}{4}$$

$$P(A_3) = P(\{\omega | X(\omega) = 2\}) = f_X(2) = \tfrac{6}{16} = \tfrac{3}{8}$$

$$P(A_4) = P(\{\omega | X(\omega) = 3\}) = f_X(3) = \tfrac{4}{16} = \tfrac{1}{4}$$

and

$$P(A_5) = P(\{\omega | X(\omega) = 4\}) = f_X(4) = \tfrac{1}{16}$$

That is, the probability distribution of X is given by $f_X(0) = f_X(4) = \frac{1}{16}$, $f_X(1) = f_X(3) = \frac{1}{4}$, and $f_X(2) = \frac{3}{8}$. We considered also the random variables $2X + 3$ and $|X - 2|$ and found the partitions $\mathfrak{A}_{2X+3}$ and $\mathfrak{A}_{|X-2|}$. In a straightforward way, we find the probability distribution of $2X + 3$ to be given by $f_{2X+3}(3) = f_{2X+3}(11) = \frac{1}{16}$, $f_{2X+3}(5) = f_{2X+3}(9) = \frac{1}{4}$, and $f_{2X+3}(7) = \frac{3}{8}$, and we find the probability distribution of $|X - 2|$ to be given by $f_{|X-2|}(0) = \frac{3}{8}$, $f_{|X-2|}(1) = \frac{1}{2}$, and $f_{|X-2|}(2) = \frac{1}{8}$. Let us also consider the random variable $4 - X$, which is just the number of tails in four tosses of a coin. In fact, $f_{4-X}(0) = f_{4-X}(4) = \frac{1}{16}$, $f_{4-X}(1) = f_{4-X}(3) = \frac{1}{4}$, and $f_{4-X}(2) = \frac{3}{8}$. This example shows that two *distinct* random variables (X and $4 - X$ in this case) may have the *same* probability distribution.

Consider now two random variables X and Y on a sample space Ω. Let $x_1, x_2, \ldots, x_k$ be the values of X on the events $A_1, A_2, \ldots, A_k$, respectively, and let $y_1, y_2, \ldots, y_m$ be the values of Y on the events $B_1, B_2, \ldots, B_m$, respectively.

Definition 5.4 The function $f_{X,Y}$ defined on the set of ordered pairs $\{(x_i, y_j), 1 \leq i \leq k, 1 \leq j \leq m\}$ by $f_{X,Y}(x_i, y_j) = P(A_i \cap B_j), 1 \leq i \leq k, 1 \leq j \leq m$, is called the *joint probability distribution* of X and Y.

To specify the joint probability distribution of X and Y, we must specify the pairs of values (x_i, y_j) together with the probabilities $P(A_i \cap B_j) = P(\{\omega | X(\omega) = x_i \text{ and } Y(\omega) = y_j\}), 1 \leq i \leq k, 1 \leq j \leq m$. $f_{X,Y}(x_i, y_j)$ is the probability that $X = x_i$ *and* $Y = y_j$; hence, $f_{X,Y}(x_i, y_j) \geq 0$. Moreover, we know that $A_i = \bigcup_{j=1}^{m} (A_i \cap B_j)$ and that this union is a union of mutually exclusive events. Therefore, for each $i = 1, 2, \ldots, k$,

$$f_X(x_i) = P(A_i) = \sum_{j=1}^{m} P(A_i \cap B_j) = \sum_{j=1}^{m} f_{X,Y}(x_i, y_j) \tag{5.7}$$

In the same way, $B_j = \bigcup_{i=1}^{k} (A_i \cap B_j)$ and for each $j = 1, 2, \ldots, m$,

$$f_Y(y_j) = P(B_j) = \sum_{i=1}^{k} P(A_i \cap B_j) = \sum_{i=1}^{k} f_{X,Y}(x_i, y_j) \tag{5.8}$$

Thus, the probability distributions f_X and f_Y of X and Y can be obtained from the joint probability distribution $f_{X,Y}$ of X and Y.

Joint probability distributions of two random variables are best exhibited by a double-entry table similar to the one in Figure 5.3. In particular, in Figure 5.4 we have duplicated Figure 5.3, replacing the

$X \backslash Y$	y_1	y_2	$\ldots$	y_m	f_X
x_1	$f_{X,Y}(x_1, y_1)$	$f_{X,Y}(x_1, y_2)$	$\ldots$	$f_{X,Y}(x_1, y_m)$	$f_X(x_1)$
x_2	$f_{X,Y}(x_2, y_1)$	$f_{X,Y}(x_2, y_2)$	$\ldots$	$f_{X,Y}(x_2, y_m)$	$f_X(x_2)$
$\cdot$	$\cdot$				
$\cdot$	$\cdot$				
$\cdot$	$\cdot$				
x_k	$f_{X,Y}(x_k, y_1)$	$f_{X,Y}(x_k, y_2)$	$\ldots$	$f_{X,Y}(x_k, y_m)$	$f_X(x_k)$
f_Y	$f_Y(y_1)$	$f_Y(y_2)$	$\ldots$	$f_Y(y_m)$	1

FIGURE 5.4

events of Figure 5.3 by their probabilities. We have observed in Eqs. (5.7) and (5.8) that the sum across the row in Figure 5.4 corresponding to x_i gives the entry in the same row of the last column and that the sum down the column corresponding to y_j gives the entry in the same column of the last row. Further, since f_X and f_Y are probability distributions, the sum across the last row is 1, as is the sum down the last column. In statistical terminology, the last column and the last row are called, respectively, the *marginal probability distribution* of X and the *marginal probability distribution* of Y. This terminology refers to the positions of this row and column on the margin. In point of fact, one ought to omit the word "marginal" and call them by their proper names, viz., the probability distribution of X and the probability distribution of Y.

Example 5.7

We return to Example 5.5. Ω is a sample space of descriptions of all five-card hands drawn from a standard deck, X is the number of black aces in the hand, and Y is the number of red cards in the hand. Suppose P is the equally likely measure on the events of Ω. We have previously noted that X takes the values 0, 1, and 2, while Y takes the values 0, 1, 2, 3, 4, and 5. In order to write out the joint probability distribution of X and Y, we must evaluate $P(\{\omega | X(\omega) = i \text{ and } Y(\omega) = j\})$ for $i = 0, 1, 2, j = 0, 1, 2, 3, 4, 5$. The results are given in Figure 5.5, where we have set $C = 1/\binom{52}{5}$. The zeros near the lower right corner of the table in Figure 5.5 reflect the fact that the hand cannot contain more than five cards. Consider now two other random variables in the same context, $5 - Y$ and Z, the number of red queens in the hand. $5 - Y$ may be seen to be

X \ Y	0	1	2
0	$C\binom{24}{5}$	$C\binom{26}{1}\binom{24}{4}$	$C\binom{26}{2}\binom{24}{3}$
1	$C\binom{2}{1}\binom{24}{4}$	$C\binom{2}{1}\binom{26}{1}\binom{24}{3}$	$C\binom{2}{1}\binom{26}{2}\binom{24}{2}$
2	$C\binom{24}{3}$	$C\binom{26}{1}\binom{24}{2}$	$C\binom{26}{2}\binom{24}{1}$
f_Y	$C\binom{26}{5}$	$C\binom{26}{1}\binom{26}{4}$	$C\binom{26}{2}\binom{26}{3}$

FIGURE 5.5

the number of black cards in the hand. It is easily checked that the joint probability distribution of Z and $5 - Y$ is the same as that of X and Y. *X and Y are different random variables from Z and $5 - Y$, yet each pair has the same joint probability distribution.*

Finally, let $X_1, X_2, \ldots, X_r$ be $r \geq 2$ random variables on a sample space Ω. Let us consider the set, say V, of all ordered r-tuples $(x_1, x_2, \ldots, x_r)$ of numbers for which X_i take the value x_i for each of $i = 1, 2, \ldots, r$.

Definition 5.5 The function $f_{X_1, X_2, \ldots, X_r}$ defined on the set V by

$$f_{X_1,X_2,\ldots,X_r}(x_1, x_2, \ldots, x_r) = P(\{\omega | X_1(\omega) = x_1, X_2(\omega) = x_2, \ldots, X_r(\omega) = x_r\})$$

is called the *joint probability distribution* of $X_1, X_2, \ldots, X_r$.

It is evident that Definition 5.5 is just an extension of Definition 5.4 to the case in which there are more than two random variables under consideration. In particular, analogues to Eqs. (5.7) and (5.8) hold when $r > 2$. Since our requirements in the way of joint probability distributions of more than two random variables are modest and the notational difficulties are not, we defer the properties of joint probability distributions to certain optional exercises.

EXERCISES 5.4

1. In the context of Example 5.6, find the probability distribution of the following random variables.

(**a**) The length Z of the longest block of consecutive tails in ω ($Z = 0$ if no tails are tossed).

3	4	5	f_X
$C\binom{26}{3}\binom{24}{2}$	$C\binom{26}{4}\binom{24}{1}$	$C\binom{26}{5}$	$C\binom{50}{5}$
$C\binom{2}{1}\binom{26}{3}\binom{24}{1}$	$C\binom{2}{1}\binom{26}{4}$	0	$C\binom{2}{1}\binom{50}{4}$
$C\binom{26}{3}$	0	0	$C\binom{50}{3}$
$C\binom{26}{3}\binom{26}{2}$	$C\binom{26}{4}\binom{26}{1}$	$C\binom{26}{5}$	1

(**b**) The number W of the toss on which the last tail appears ($W = 0$ if no tails are tossed).
(**c**) $W - Z$.

2. In the context of Example 5.6, find the joint distribution of X and the random variable W of Exercise 1. Exhibit the results in the form of a double-entry table and verify Eqs. (5.7) and (5.8).

3. An urn contains eight balls of which three are red and five are black. Four balls are drawn from the urn without replacement in such a way that each combination of balls is equally likely. What is the probability distribution of X, the number of red balls drawn? What is the joint probability distribution of X and the number Y of black balls drawn?

4. Three balls are placed successively in one of three boxes numbered 1, 2, and 3, and all possible $3^3 = 27$ placements are regarded as equally likely. Find the probability distributions of the number X_1 of balls in box number 1 and of the number X_2 of balls in box number 2. Find the joint probability distribution of X_1 and X_2.

5. Suppose four cards are dealt from a standard deck and all $\binom{52}{4}$ hands are regarded as equally likely. Write out in double-entry form the joint probability distribution of the number X of aces dealt and the number Y of spades dealt.

6. Suppose P is the equally likely measure on $\Omega = \{\omega_1, \omega_2, \ldots, \omega_n\}$. Give the form of the probability distribution of I_A for an event A of Ω. Give the form of the joint probability distribution of I_A and I_B for events A and B of Ω.

7. Suppose X is a random variable on Ω that assumes the distinctive values $x_1, x_2, \ldots, x_k$ on the events $A_1, A_2, \ldots, A_k$, respectively. If P is a probability measure on the events of Ω, we have called the function f_X given by $f_X(x_i) = P(A_i)$, $i = 1, 2, \ldots, k$, the probability distribution of X. Given the occurrence of an event A of Ω, then P_A is another probability measure on the events of Ω. It is now customary to call the probability distribution of X, the *conditional probability distribution* of X given A. We can denote this by $f_{X|A}$, and it is the function defined by $f_{X|A}(x_i) = P_A(A_i)$, $i = 1, 2, \ldots, k$. In the context of Example 5.6, find the conditional probability distribution of X given (i) that the first toss is heads and (ii) that the first toss is tails.

8. Is it generally true that $f_{X,Y} = f_{Y,X}$? If not, what may be said of X and Y when $f_{X,Y} = f_{Y,X}$. Can you give a simple example of two random variables for which $f_{X,Y} = f_{Y,X}$?

9. Let X be a random variable with a probability distribution given by

x	-2	-1	0	1	2	3	7	12
$f_X(x)$	.1	.15	.2	.2	.15	.1	.05	.05

What is the probability of the event that
(a) X is even.
(b) X is negative.
(c) X is less than -1 and odd.
(d) X takes a value between 1 and 8 inclusive.
(e) X takes a value between 1 and 7 inclusive.

10. Let X and Y be two random variables with a joint probability distribution given by

$X \backslash Y$	-1	1	2	5
-1	$\frac{1}{27}$	$\frac{1}{9}$	$\frac{1}{9}$	$\frac{1}{27}$
1	$\frac{1}{9}$	$\frac{2}{9}$	$\frac{1}{9}$	0
5	$\frac{4}{27}$	$\frac{1}{9}$	0	0

What is the probability of the event that
(a) Y is even.
(b) Y is even and X^2 is less than 2.
(c) XY is odd.
(d) X is positive but less than 5.
(e) Y is positive but less than 5.
Find the probability distributions of $X + Y$ and XY. Find the conditional probability that Y is odd given that X is negative.

***11.** Find the joint probability distribution of $X_1, X_2, \ldots, X_{r-1}$ in terms of the joint probability distribution of $X_1, X_2, \ldots, X_r$; i.e., obtain a version of Eq. (5.6) in this more general case. Generalize this result by induction.

***12.** Suppose that X and Y are two random variables and that each is a function of the other. What can be said of f_X and f_Y? What can be said of $f_{X,Y}$?

***13.** Let X be a random variable on Ω that assumes the values $x_1, x_2, \ldots, x_k$, and suppose $h(X)$ assumes the values $h_1, h_2, \ldots, h_q$. Show that the probability distribution of $h(X)$ is determined by the probability distribution of X, by showing that

$$f_{h(X)}(h_j) = \sum_{\{x_i | h(x_i) = h_j\}} f_X(x_i) \qquad j = 1, 2, \ldots, q$$

***14.** Let X and Y be two random variables on Ω assuming the values $x_1, x_2, \ldots, x_k$ and $y_1, y_2, \ldots, y_m$, respectively. Suppose $h(X, Y)$ assumes the values $h_1, h_2, \ldots, h_q$. Show that the probability distribution of $h(X, Y)$ is determined by the joint probability distribution of X and Y, by showing that

$$f_{h(X,Y)}(h_q) = \sum_{\{(x_i,y_j) | h(x_i,y_j) = h_q\}} f_{X,Y}(x_i, y_i)$$

***15.** State the meaning of the equations in the last two exercises, and then give a general rule for determining the probability distribution of $h(X_1, X_2, \ldots, X_r)$ in terms of the joint probability distribution of $X_1, X_2, \ldots, X_r$.

5.5 Independence of random variables

We take as given a sample space Ω and a probability measure P on Ω, and we let X and Y be random variables on Ω. As we have done before, we shall write $x_1, \ldots, x_k$ for the distinct values assumed by X_j, and $y_1, \ldots, y_m$ for the distinct values assumed by Y. Further, A_i will stand for the event $X = x_i$, $i = 1, \ldots, k$, and B_j for the event $Y = y_j$, $j = 1, \ldots, m$.

The notion of independent random variables is quite easily described intuitively. We say that X and Y are independent if, roughly speaking, a knowledge that X has assumed a given value, say x_i, does not effect the probability that Y will assume any given value, say y_j. More precisely, we may say that X and Y are independent random variables if for each $i = 1, \ldots, k$ the event A_i (i.e., the event $X = x_i$) is independent of each of the events B_j (i.e., the event $Y = y_j$) $j = 1, \ldots, m$. Now the events $A_1, \ldots, A_k$ comprise the partition $\mathcal{A}_X$, and the events $B_1, \ldots, B_m$ comprise the partition $\mathcal{A}_Y$. Accordingly, we make the following definition.

Definition 5.6 If P is a probability measure on the events of Ω and if X and Y are random variables on Ω, then X and Y are said to be *independent random variables* if the partitions $\mathcal{A}_X$ and $\mathcal{A}_Y$ are independent.

Independence of partitions $\mathcal{A}_X$ and $\mathcal{A}_Y$, it may be recalled, means that any event from $\mathcal{A}_X$ is independent of any event from $\mathcal{A}_Y$. We have taken $\mathcal{A}_X = \{A_1, A_2, \ldots, A_k\}$ and $\mathcal{A}_Y = \{B_1, B_2, \ldots, B_m\}$. Therefore, if X and Y are independent, then

$$P(A_i \cap B_j) = P(A_i)P(B_j), \qquad 1 \le i \le k, \quad 1 \le j \le m \tag{5.9}$$

Two remarks should be made in connection with Definition 5.6. First, the independence of X and Y is a property defined with reference to a definite probability measure P on Ω. Thus, we ought to say X and Y are independent with respect to P. However, the measure P that is intended is usually clear from the context. Second, the conditions of independence as given in Eq. (5.9) can be phrased in terms of the joint probability distribution $f_{X,Y}$ of X and Y and the probability distributions f_X and f_Y or X and Y. Indeed, $P(A_i \cap B_j) = f_{X,Y}(x_i, y_j)$, $P(A_i) = f_X(x_i)$, and $P(B_j) = f_Y(y_j)$. Therefore, if X and Y are independent,

$$f_{XY}(x_i, y_j) = f_X(x_i)f_Y(y_j), \qquad 1 \leq i \leq k, \quad 1 \leq j \leq m \tag{5.10}$$

We often refer to Eq. (5.10) in the following loose way: X and Y are independent if the joint probability distribution of X and Y is the product of the probability distribution of X with the probability distribution of Y.

It might also be stressed here that we have accomplished our purpose with Definition 5.6. That is, if X and Y are independent, then a knowledge that $X = x_i$, say, does not change the probability distribution of Y. Specifically, for each i with $P(A_i) > 0$,

$$P_{A_i}(\{\omega \mid Y(\omega) = j\}) = P_{A_i}(B_j) = P(B_j) = P(\{\omega \mid Y(\omega) = y_j\})$$

for every $j = 1, 2, \ldots, m$, by the assumed independence of $\mathcal{A}_X$ and $\mathcal{A}_Y$.

The notion of independence can be carried over to more than two random variables in the following way.

Definition 5.7 If P is a probability measure on the events of Ω and if $X_1, X_2, \ldots, X_r$ are $r \geq 2$ random variables on Ω, $X_1, X_2, \ldots, X_r$ are said to be *mutually independent random variables* if the partitions $\mathcal{A}_{X_1}$, $\mathcal{A}_{X_2}, \ldots, \mathcal{A}_{X_r}$ are mutually independent partitions of Ω.

The mutual independence of partitions $\mathcal{A}_{X_1}, \mathcal{A}_{X_2}, \ldots, \mathcal{A}_{X_r}$ means again that for any event A_1 from $\mathcal{A}_{X_1}$, any event A_2 from $\mathcal{A}_{X_2}, \ldots$, any event A_r from $\mathcal{A}_{X_r}$, the events $A_1, A_2, \ldots, A_r$ are mutually independent.

Another way of stating the conditions of mutual independence of X_1, $X_2, \ldots, X_r$ is as follows: for any value x_1 assumed by X_1, any value x_2 assumed by $X_2, \ldots$, any value x_r assumed by X_r,

$$P(\{\omega \mid X_1(\omega) = x_1, X_2(\omega) = x_2, \ldots, X_r(\omega) = x_r\}) = \prod_{i=1}^{r} P(\{\omega \mid X_i(\omega) = x_i\}) \tag{5.11}$$

Rewriting Eq. (5.11) in terms of probability distributions, we find that these conditions are

$$f_{X_1, X_2, \ldots, X_r}(x_1, x_2, \ldots, x_r) = \prod_{i=1}^{r} f_{X_i}(x_i) \tag{5.12}$$

That is, the joint probability distribution of $X_1, X_2, \ldots, X_r$ is the product of the individual probability distributions of $X_1, X_2, \ldots, X_r$.

Remarks: The verification of the mutual independence of random variables is ultimately the verification of the mutual independence of events and hence the computations of specific probabilities. This may be straightforward but, in practice, may be prohibitively tedious. For example, if we are considering five random variables each taking ten values, there are 10^5 relations like Eq. (5.12) to be checked (some of these are redundant in fact, but few relative to 10^5). As in the case of the independence of events, it is somewhat easier to find dependent random variables than independent ones. For instance, it is clear in Example 5.7 that X and Y are not independent inasmuch as some values of X are incompatible with certain values of Y. Many other examples of the same sort may be constructed. In the direction of independence, it is to be noted that whenever we find independent partitions, *any* random variables giving rise to these partitions are independent regardless of the particular values they assume. Thus, the reader can find mutually independent random variables by referring to the mutually independent partitions given in Section 4.4.

EXERCISES 5.5

1. Let the sample space Ω for throwing a die twice be the Cartesian product of the set $\{1, 2, 3, 4, 5, 6\}$ with itself. Let P be the equally likely measure on the events of Ω. Show directly that the score X on the first throw and the score Y on the second throw are independent random variables. Are the random variables $2X$ and Y independent? Are the random variables $2X$ and $3Y$ independent? Are the random variables X and $X + Y$ independent?

2. Suppose that P is a probability measure on the events of a sample space Ω and that A and B are independent events of Ω. Find two pairs of independent random variables on Ω.

3. Suppose that P is a probability measure on the events of a sample space Ω and that $A_1, A_2, \ldots, A_r$ are mutually independent events of Ω. Find two sets of r random variables on Ω that are mutually independent.

4. Show that Eq. (5.12) follows from the conditions of Definition 5.7.

5. Consider the experiment of drawing a 13-card hand from a standard deck. Suppose all $\binom{52}{13}$ hands are regarded as equally likely. Explain why you expect to find *dependence* of the following random variables.

(a) The number X of spades drawn and the number Y of black cards drawn.
(b) X and the number Z of red cards drawn.
(c) X and the number W of aces drawn.
(d) Y and W.

In each case, give an event A from one partition and an event B from the other that can easily be shown to be dependent.

6. Suppose P is a probability measure on the events of Ω. Show that two random variables X and Y on Ω are independent if and only if X and $2Y$ are independent.

7. Suppose P is a probability measure on the events of Ω. Show that any random variable X on Ω is independent of I_Ω.

8. Suppose that P is a probability measure on the events of Ω and that X and Y are independent random variables on Ω. Show that $A = \{\omega | X(\omega) \leq 0\}$ and $B = \{\omega | Y(\omega) \leq 0\}$ are independent events of Ω. Show that I_A and I_B are independent random variables on Ω.

9. Suppose P is a probability measure on the events of Ω. Random variables $X_1, X_2, \ldots, X_r$ are called *pairwise independent* if X_i and X_j are independent for all choices of i and j with $i \neq j$. Show that if $X_1, X_2, \ldots, X_r$ are mutually independent random variables, then they are pairwise independent (see Exercise 7 of Section 4.5). Show, by example, that the converse to this is not true.

***10.** Show that the conditions of Definition 5.7 follow from Eq. (5.12). (See Exercise 11 of Section 5.4 and use induction on the number r of random variables.)

***11.** Suppose that P is a probability measure on the events of Ω and that X and Y are independent random variables on Ω. Show that $g(X)$ and Y are independent random variables on Ω. (See Exercise 7 of Section 5.3 and Exercise 12 of Section 4.5.)

CHAPTER SIX

Independent Trials

6.1 Introduction

The topics covered in this chapter have many important applications in a variety of disciplines. Since we can no more than hint at them all, exercise your imagination while perusing the text and the problems.

Specifically, we treat the notion of independent trials of an experiment through three sections. During this development, we discuss two important probability distributions—the binomial and the multinomial. In the final section of the chapter, we discuss a model called random sampling without replacement and treat a third distribution—the hypergeometric.

6.2 Independent trials

Consider an experiment having the following description: a given experiment, call it $\mathcal{E}$, is performed n times in succession. This composite experiment will be denoted by $\mathcal{E}^n$, and we will say $\mathcal{E}^n$ *is the experiment consisting of n trials* (*or replications*) *of the experiment* $\mathcal{E}$. We have discussed such composite experiments previously, but without particular stress on their special structure. In this section, however, we will emphasize this structure to the following extent: a probability measure on a sample space is found for the experiment $\mathcal{E}^n$, which embodies the intuitive notion that *the n trials of* $\mathcal{E}$ *are performed independently of each other under identical conditions.* Before we determine such a probability measure, we must translate into mathematical terms the intuitive notions embodied by the phrases "performed independently" and "under identical conditions."

Example 6.1

Coin Tossing Let us suppose $\mathcal{E}$ is the experiment of tossing a coin and, therefore, that $\mathcal{E}^n$ is the experiment of tossing a coin n times in succession. Further, let us assume that the trials of $\mathcal{E}$ are performed under the same conditions. Therefore, we can initially agree that in each trial of $\mathcal{E}$ either heads or tails appears and that the probability of heads appearing is $\frac{1}{2}$. In other words, let us adopt the sample space $\Omega = \{H, T\}$ together with a probability measure P on Ω determined by $P(\{H\}) = \frac{1}{2}$ for a *single* replication of $\mathcal{E}$. Having done this, we may describe an outcome of the experiment $\mathcal{E}^n$ as a sequence of heads and tails of length n, and we find that a natural choice of sample space is Ω^n, the Cartesian product of the set $\{H, T\}$ with itself n times. For a given sample point

$$\omega = (\omega_1, \omega_2, \ldots, \omega_n) \in \Omega^n$$

ω_i (which is either H or T) is the outcome of the ith trial of $\mathcal{E}$.

A good choice of probability measure on Ω^n for the experiment $\mathcal{E}^n$ is much more delicate. We have in effect agreed that each of the events $A_i = \{\omega |\ \omega \text{ has H in the } i\text{th position}\}$ of Ω^n, $i = 1, 2, \ldots, n$, ought to be assigned probability $\frac{1}{2}$. That is, the probability of heads on the ith trial is $\frac{1}{2}$. There are, however, many different probability measures on Ω^n that give the events A_i probability $\frac{1}{2}$. How should we choose one of these? We note (and show below) that a single probability measure would be singled out *if it were further agreed that the events* $A_1, A_2, \ldots, A_n$ *should be mutually independent with respect to the probability measure chosen.* Is such an assumption of mutual independence of events justified? Let us recall from our previous discussion of independence and dependence of events in Section 4.4 that if $A_1, A_2, \ldots, A_n$ are *not* mutually independent, then a knowledge that certain of these events have occurred will have an influence on the probability of occurrence of others. For example, we might then find that the occurrence of A_1, A_2, and A_3 gives A_4 a conditional probability of $\frac{1}{3}$, not $\frac{1}{2}$. We cannot argue that this does not happen in some real coin-tossing experiment. What we can and do argue is that a model in which this type of phenomenon is not allowed to happen is worthy of further study, and that it is a model which, on an empirical basis, accords well with experience. As we shall see, there is one probability measure on Ω^n that assigns each of $A_1, A_2, \ldots, A_n$ probability $\frac{1}{2}$ and at the same time makes these events mutually independent.

Example 6.2

Quality Control Consider a production process that produces items successively. Since no production process is perfect, we expect that even with the best possible care some defective items will be produced. Let

$\mathcal{E}$ be the experiment of inspecting an item and classifying it as defective or nondefective. $\mathcal{E}^n$ consists then of n successive inspections and classifications. Let us agree, for a single replication of $\mathcal{E}$, that the sample space is $\Omega = \{\text{defective, nondefective}\}$ together with a probability measure P on Ω determined by $P(\{\text{defective}\}) = p$, $0 < p < 1$. It would again be natural to adopt Ω^n, the Cartesian product of Ω with itself n times, as a sample space for $\mathcal{E}^n$. It would further seem appropriate, since we suppose the trials of $\mathcal{E}$ to be held under the same conditions, to assign probability p to each event of Ω^n of the form $A_i = \{\omega | \omega \text{ has defective in the } i\text{th position}\}$, $i = 1, 2, \ldots, n$, the event that the ith trial of $\mathcal{E}$ results in a defective item. Once again there are various probability measures on Ω^n under which each A_i has probability p. There is, however, only one that at the same time makes the events $A_1, A_2, \ldots, A_n$ mutually independent. If we adopt it for the experiment $\mathcal{E}^n$, we are saying, in effect, that the factors causing a given item to be defective are really "chance" factors that do not persist systematically and thereby cause other items to be defective also.

Example 6.3

Roulette Let $\mathcal{E}$ be the experiment of spinning a roulette wheel and thereby placing a ball in one of 38 compartments numbered 00, 0, 1, 2, . . . , 36. $\mathcal{E}^n$ is the experiment of n successive plays of a roulette wheel. For a single replication of $\mathcal{E}$, let us take P to be the equally likely measure on the events of $\Omega = \{00, 0, 1, 2, \ldots, 36\}$. For $\mathcal{E}^n$ let us adopt the sample space Ω^n, with the obvious interpretation being given to each sample point of Ω^n. Consider the events of Ω^n of the form $A_{i,j} = \{\omega | \omega \text{ has } j \text{ in the } i\text{th position}\}$, $i = 1, 2, \ldots, n$, $j = 00, 0, 1, 2, \ldots, 36$, i.e., $A_{i,j}$ is the event that the ith play results in j. Let us set the probability of each of these events $= \frac{1}{38}$. We subsequently could select a probability measure on Ω^n by asking that the partitions $\mathcal{A}_1 = \{A_{1,00}, A_{1,0}, A_{1,1}, \ldots, A_{1,36}\}$, $\mathcal{A}_2 = \{A_{2,00}, A_{2,0}, A_{2,1}, \ldots, A_{2,36}\}, \ldots, \mathcal{A}_n = \{A_{n,00}, A_{n,0}, A_{n,1}, \ldots, A_{n,36}\}$ be mutually independent partitions of Ω^n with respect to it.

If, rather, a probability measure were chosen without this property, we would be saying that the outcomes of certain plays of the wheel have a positive or negative effect on the probabilities of certain outcomes of other plays. In other words, we would be saying that the wheel has a kind of memory. This supposition does have its adherents, and they may be found at almost any roulette table in Las Vegas recording the sequence of plays with the conviction that, for example, a dearth of 13's in 100 plays gives 13 a probability larger than $\frac{1}{38}$ on the 101st play. We, however, will study the probability measure that arises by supposing $\mathcal{A}_1, \mathcal{A}_2, \ldots, \mathcal{A}_n$ are mutually independent. (The apparent difference in assuming mutual independence of partitions here and mutual independence of events in Examples 6.1 and 6.2 is illusory. The mutual independence of the events $A_1, A_2, \ldots, A_n$ in the first two examples is really equivalent to the mutual independence of the partitions $\mathcal{A}_1 = \{A_1, A_1'\}$, $\mathcal{A}_2 = \{A_2, A_2'\}, \ldots, \mathcal{A}_n = \{A_n, A_n'\}$ as was pointed out in Section 4.4.)

We now proceed to construct the probability measure that embodies the mutual independence described in these examples. Suppose, for the experiment $\mathcal{E}$, that Ω is a sample space and that P is a probability measure on the events of Ω. For the experiment $\mathcal{E}^n$, we take the sample space Ω^n, the Cartesian product of Ω with itself n times. We write the sample points of Ω^n as $\omega = (\omega_1, \omega_2, \ldots, \omega_n)$, wherein ω_i denotes the outcome of the ith trial of $\mathcal{E}$. Of course, $\omega_i \in \Omega$ for each of $i = 1, 2, \ldots, n$. (Note that we have made a notational change and that we have avoided listing the members of Ω.)

To define a probability measure, say $\bar{P}$, on Ω^n, it would suffice to define $\bar{P}(\{\omega\})$ for each sample point ω of Ω^n. Suppose $\omega = (\omega_1, \omega_2, \ldots, \omega_n)$, and let A_i be the event that the ith trial of $\mathcal{E}$ results in ω_i, $i = 1, 2, \ldots, n$. Then

$$\{\omega\} = A_1 \cap A_2 \cap \cdots \cap A_n \tag{6.1}$$

Now, the event A_i specifies the outcome of the ith replica of $\mathcal{E}$, and so collectively, the events A_i specify the outcomes of the n trials of $\mathcal{E}$. If we ask that these events be mutually independent with respect to $\bar{P}$, then it must be that

$$\bar{P}(\{\omega\}) = \bar{P}(A_1 \cap A_2 \cap \cdots \cap A_n) = \prod_{i=1}^{n} \bar{P}(A_i) \tag{6.2}$$

If further, each trial of $\mathcal{E}$ is run under the same conditions and according to the probability measure P on Ω, then A_i, being the event that the ith trial results in ω_i, should be assigned the probability $P(\{\omega_i\})$, i.e., $\bar{P}(A_i) = P(\{\omega_i\})$, $i = 1, 2, \ldots, n$. In conjunction with Eq. (6.2) this leads us to an assignment of

$$\bar{P}(\{\omega\}) = \bar{P}(\{(\omega_1, \omega_2, \ldots, \omega_n)\}) = \prod_{i=1}^{n} P(\{\omega_i\}) \tag{6.3}$$

In other words, we contemplate assigning to each simple event of Ω^n a probability equal to the product of the probabilities of the individual outcomes at each trial. This, as we know from Section 3.2, will define a probability measure $\bar{P}$ on Ω^n if the sum of these "probabilities" is 1. We prove that this is true in the following theorem.

Theorem 6.1 Let Ω be a sample space and P be a probability measure on Ω. For a fixed positive integer n, let Ω^n be the Cartesian product of Ω with itself n times. For $\omega = (\omega_1, \omega_2, \ldots, \omega_n) \in \Omega^n$, define $\bar{P}$ at $\{\omega\}$ by Eq. (6.3) and for any event A of Ω^n, let $\bar{P}(A) = \sum_{\omega \in A} \bar{P}(\{\omega\})$. Then $\bar{P}$ is a probability measure on Ω^n.

PROOF: It suffices to show that $\bar{P}$ as it is defined on Ω^n satisfies $\bar{P}(\Omega^n) = \sum_{\omega \in \Omega^n} \bar{P}(\{\omega\}) = 1$ and we do this by induction on n. For this purpose, we denote $\bar{P}$ on the events of Ω^n by P^n. If $n = 1$, P^1 reduces to P on the events of $\Omega^1 = \Omega$, so the result is true. Suppose that it is true for $n = k$, i.e., that

$$P^k(\Omega^k) = \sum_{\omega \in \Omega^k} P^k(\{\omega\}) = 1$$

For $n = k + 1$ and $\omega = (\omega_1, \omega_2, \ldots, \omega_{k+1})$, we see that

$$\begin{aligned} P^{k+1}(\{\omega\}) &= P^{k+1}(\{(\omega_1, \omega_2, \ldots, \omega_{k+1})\}) \\ &= P(\{\omega_1\})P(\{\omega_2\}) \cdots P(\{\omega_k\})P(\{\omega_{k+1}\}) \\ &= P^k(\{(\omega_1, \omega_2, \ldots, \omega_k)\})P(\{\omega_{k+1}\}) \end{aligned}$$

To perform the sum $\sum_{\omega \in \Omega^{k+1}} P^{k+1}(\{\omega\})$, we first sum over all $\omega_{k+1} \in \Omega$ holding $(\omega_1, \omega_2, \ldots, \omega_k) \in \Omega^k$ fixed.

$$\begin{aligned} \sum_{\omega_{k+1} \in \Omega} P^{k+1}(\{(\omega_1, \omega_2, \ldots, \omega_{k+1})\}) &= \sum_{\omega_{k+1} \in \Omega} P^k(\{(\omega_1, \omega_2, \ldots, \omega_k)\})P(\{\omega_{k+1}\}) \\ &= P^k(\{(\omega_1, \omega_2, \ldots, \omega_k)\}) \end{aligned}$$

since $\sum_{\omega_{k+1} \in \Omega} P(\{\omega_{k+1}\}) = 1$. Then summing $P^k(\{(\omega_1, \omega_2, \ldots, \omega_k)\})$ over all choices of $(\omega_1, \omega_2, \ldots, \omega_k) \in \Omega^k$, we find that the result follows by the induction hypothesis.

The measure $\bar{P}$ defined on Ω^n as in the statement of Theorem 6.2 is called a *product probability measure on* Ω^n because of Eq. (6.3). It is often denoted by P^n. Beginning with a sample space Ω and a probability measure P on Ω for the experiment $\mathcal{E}$, P^n *is the choice of probability measure on* Ω^n *implied by the following statement: let* $\mathcal{E}^n$ *be the experiment of* n *independent trials of* $\mathcal{E}$.

In point of fact, it has not yet been established that the mutual independence of events like those given in Examples 6.1, 6.2, and 6.3 actually holds with respect to P^n. More specifically, let $\omega = (\omega_1, \omega_2, \ldots, \omega_n) \in \Omega^n$, and consider the events $A_1, A_2, \ldots, A_n$ of Ω^n, where A_i is the event that the ith trial of $\mathcal{E}$ results in ω_i, $i = 1, 2, \ldots, n$. To obtain Eq. (6.2), we argued that if $A_1, A_2, \ldots, A_n$ were to be mutually independent with respect to P^n, then it would be necessary to have

$$P^n(A_1 \cap A_2 \cap \cdots \cap A_n) = \prod_{i=1}^{n} P^n(A_i)$$

That the events $A_1, A_2, \ldots, A_n$ do turn out to be mutually independent with respect to P^n is already implicit in previous exercises but will be repeated in this section in a more explicit fashion. This mutual independence of events will also carry over to the mutual independence of any random variables $X_1, X_2, \ldots, X_n$ that are such that the value of X_i at each $\omega = (\omega_1, \omega_2, \ldots, \omega_n)$ depends only on ω_i, $i = 1, 2, \ldots, n$. In other words, the value of X_i is determined by the outcome of the ith trial of $\mathcal{E}$, $i = 1, 2, \ldots, n$. Some further exercises at the end of the section establish this fact.

Finally, we notice that P^n is the only probability measure on Ω^n that satisfies Eqs. (6.2) and (6.3). Any other probability measure sharing these properties would have to assign each simple event of Ω^n the same probability as does P^n. But then it must be equal to P^n.

EXERCISES 6.2

1. Describe five realistic situations that may be described as performing independent trials of an experiment under identical conditions.

2. Describe five realistic situations where the assumption that the trials of an experiment are independent is *not* justified, and state the reasons for your answer.

3. Let a coin be tossed four times. Let Ω be the sample space $\{H, T\}$ for one toss, and let $P(\{H\}) = \frac{1}{2}$, $P(\{T\}) = \frac{1}{2}$. Write out the product probability measure P^4 on Ω^4, and check that P^4 is the equally likely measure on Ω^4. Could you have predicted this without making the computation?

4. Continue with Ω^4 and P^4 as in Exercise 3. In each of the following, guess whether the events A and B described there are independent or not. Having made the guess, verify the guess by computing $P(A \cap B)$, $P(A)$, $P(B)$.

(a) A: the event that the first toss results in heads.
B: the event that the third toss results in heads.

(b) A: the event that the second toss results in tails.
B: the event that the second toss results in heads.

(c) A: the event that the second toss results in tails.
B: the event that there are no more than two heads in the four tosses.

(d) A: the event that the first toss results in tails.
B: the event that at least one of the last two tosses results in a head.

(e) A: the event that the first and third tosses both result in tails.
B: the event that the second and fourth tosses do not both result in heads.

5. A balanced die is tossed twice. Let $\Omega = \{1, 2, 3, 4, 5, 6\}$ be the sample space for one toss, and let P be the equally likely probability measure on Ω. Determine the product measure P^2 on Ω^2, and verify that it is the equally likely measure on Ω^2.

6. Let Ω be any finite sample space, and suppose that P is the equally likely probability measure on Ω. Let n be a positive integer, and let P^n be the product measure on Ω^n. Show that P^n is equally likely on Ω^n.

7. An urn contains red and white balls in the proportion 1:2. Let $\mathcal{E}$ be the experiment consisting of drawing a ball, noting its color, and returning it to the urn. Assume that at each draw, every ball is as likely to be drawn as any other, and consider $\mathcal{E}^4$, the experiment of four independent trials of $\mathcal{E}$. Give a sample space and a probability measure on it for the experiment $\mathcal{E}^4$. Is this probability measure equally likely? Why?

8. Continue with the setting of Exercise 7. In each of the following, guess whether the events A and B described are independent. Then verify your guess by computing respectively the probabilities of $A \cap B$, A, and B.

(**a**) A: the first and third balls drawn are not of the same color.
B: the second ball drawn is red.

(**b**) A: the first and third balls drawn are not of the same color.
B: the second ball drawn is the same color as the first.

(**c**) A: the first and third balls drawn are not of the same color.
B: the second ball drawn is the same color as the fourth.

(**d**) A: the third ball drawn is red.
B: at least two white balls are among the first, second, and fourth balls drawn.

(**e**) A: the third ball drawn is white.
B: at most two red balls are drawn in the four draws.

***9.** Prove the following converse to the statement of Exercise 6. Let Ω be a sample space. If P^n is the product probability measure on Ω^n of a probability measure P on Ω, then P^n equally likely on Ω^n implies P is equally likely on Ω.

***10.** Let P^n be a product probability measure on Ω^n. Let $\omega \in \Omega^n$, say $\omega = (\omega_1, \omega_2, \ldots, \omega_n)$, and suppose $A_i = \{\omega \mid \omega \text{ has } \omega_i \text{ in the } i\text{th position}\}$. Prove that the events $A_1, A_2, \ldots, A_n$ are mutually independent. [Begin by showing that $P(A_1 \cap A_2 \cap \cdots \cap A_{n-1}) = P(A_1)P(A_2) \cdots P(A_{n-1})$.]

***11.** A random variable X on the sample space Ω^n for $\mathcal{E}^n$ *is said to depend only on the trials* $i_1, i_2, \ldots, i_r$ *of* $\mathcal{E}$ if X is constant on each event of the form $A = \{\omega \mid \omega \text{ has } \omega_1 \text{ in the } i_1\text{st position}, \omega_2 \text{ in the } i_2\text{nd position}, \ldots, \omega_r \text{ in the } i_r\text{th position}\}$. Let P^n be a product probability measure on Ω^n, and suppose X and Y are random variables depending on trials $1, 2, \ldots, m$ and $m + 1, m + 2, \ldots, n$, respectively. Show that X and Y are independent.

***12.** As in the previous exercise, let $X_1, X_2, \ldots, X_n$ be random variables that depend on the first, on the second, ... , on the nth trial of $\mathcal{E}$, respectively. Show that $X_1, X_2, \ldots, X_n$ are mutually independent with respect to P^n.

***13.** Suppose that n experiments $\mathcal{E}_1, \mathcal{E}_2, \ldots, \mathcal{E}_n$ are performed successively, and that the composite experiment so described is called $\mathcal{E}_1 \times \mathcal{E}_2 \times \cdots \times \mathcal{E}_n$. Find a sample space and a probability measure on it that will represent independent trials of these (possibly different) experiments when Ω_i is a sample space for $\mathcal{E}_i$ and P_i is a suitable probability measure on Ω_i, $i = 1, 2, \ldots, n$.

6.3 Bernoulli trials and the binomial distribution

Now, let us make specific the generalizations of the preceding section to an experiment $\mathcal{E}$ having just two outcomes, such as the coin tossing and the quality control examples discussed previously. Independent repeated trials of an experiment $\mathcal{E}$ having only two outcomes in a given trial are called *Bernoulli trials* after J. Bernoulli (1654–1705).

For $\mathcal{E}$, let us take a sample space $\Omega = \{S, F\}$ (S for success and F for failure) and suppose P is determined on Ω by $P(\{S\}) = p$, $0 < p < 1$. For convenience, we set $P(\{F\}) = q$ where in fact $q = 1 - p$. The setup in this case (i.e., the probability measure P^n on the events of Ω^n) is called a *Bernoulli scheme with parameters n and p.*

There are many practical situations in which one wishes to make successive trials of an experiment $\mathcal{E}$ having two outcomes. For example, insurance companies make it a practice to observe several males of a given age throughout a year to classify each as surviving at the end of the year (S) or deceased at the end of the year (F); a new drug is often tested on several patients and each is found to be cured (S) or not cured (F); each of several plantings of a strain of flowers may produce the desired color (S) or not (F), and so on.

We are going a step further in as much as we study n *independent* trials of $\mathcal{E}$. The application of this study to any practical situation is worthwhile only to the extent that one can argue that it is valid to assume such independence. This might not always be the case. For example, if an insurance company observes each of a pair of twins for survival or death in a given year, the lives of these twins may be intertwined to the extent that the event that one dies is not independent of the event that the other dies (they might be killed in the same automobile accident, for instance). Again, in an epidemic year, different people may die of the same causes. To minimize these considerations, insurance companies generally choose the subjects to be observed from different families, different locations, etc.

We assume independent trials and study P^n on Ω^n. A sample point $\omega = (\omega_1, \omega_2, \ldots, \omega_n) \in \Omega^n$ is an n-tuple with entries either S or F in each position. For any $\omega = (\omega_1, \omega_2, \ldots, \omega_n)$, we have

$$P^n(\{\omega\}) = P(\{\omega_1\})P(\{\omega_2\}) \ldots P(\{\omega_n\}) \tag{6.4}$$

Now each ω_i is either S or F so that each factor on the right-hand side of Eq. (6.4) is either $P(\{S\}) = p$ or $P(\{F\}) = q$. Then the probability of $\{\omega\}$ may be seen to depend exactly on how many of these factors are p and how many are q. In particular

$$P^n(\{\omega\}) = p^r q^{n-r} \tag{6.5}$$

if ω has r entries of S and $n - r$ entries of F. $p^r q^{n-r}$ is the probability under independent trials of obtaining success on r specified trials of $\mathcal{E}$ and $n - r$ failures on the remaining trials.

On Ω^n there is a random variable S_n that is of obvious interest. It is defined by

$$S_n(\omega) \text{ is the number of entries } S \text{ in } \omega \tag{6.6}$$

For the examples we have listed, S_n is the number of heads, the number of nondefective items, the number of survivals, the number of cures, and the number of successful plantings in n trials of $\mathcal{E}$. We will compute the probability distribution of S_n under the probability measure P^n.

To begin with, it is clear that S_n takes the values $0, 1, 2, \ldots, n$. Let

$$A_r = \{\omega |\, S_n(\omega) = r\}, \qquad r = 0, 1, 2, \ldots, n \tag{6.7}$$

$\mathcal{A}_{S_n} = \{A_0, A_1, \ldots, A_n\}$ is the partition of Ω^n induced by S_n so we must compute $P(A_r)$, $r = 0, 1, 2, \ldots, n$. Now

$$P^n(A_r) = \sum_{\omega \in A_r} P^n(\{\omega\}) \tag{6.8}$$

A sample point $\omega \in A_r$ has exactly r entries of S, and, as we have seen in Eq. (6.8), $P^n(\{\omega\})$ for such a sample point is $p^r q^{n-r}$. Accordingly, if $N(A_r)$ is the number of sample points in A_r,

$$P^n(A_r) = N(A_r) p^r q^{n-r} \tag{6.9}$$

The determination of $N(A_r)$ is a counting problem that has been previously solved. The question is, "How many n-tuples of S's and F's may be formed so that there are r entries of S and $n - r$ entries of F?" This is the number of ways r positions may be chosen for the S entries from among n, viz., $\binom{n}{r}$. Finally then,

$$P^n(A_r) = \binom{n}{r} p^r q^{n-r}, \qquad r = 0, 1, 2, \ldots, n \tag{6.10}$$

In the terminology of the probability distribution of S_n, we have found that

$$f_{S_n}(r) = \binom{n}{r} p^r q^{n-r}, \qquad r = 0, 1, 2, \ldots, n \tag{6.11}$$

This probability distribution is called the *binomial distribution,* because the right-hand side of Eq. (6.11) is the general term in the binomial expansion of $(q + p)^n$. Indeed, we have

$$(q + p)^n = \sum_{r=0}^{n} \binom{n}{r} p^r q^{n-r} \tag{6.12}$$

Now $q + p = 1$, and Eq. (6.12) says just that the sum of the probabilities that $S_n = r$ over $r = 0, 1, 2, \ldots, n$ is 1.

The random variable S_n discussed above is itself called *a binomial random variable with parameters n and p,* as is any random variable with a probability distribution given by Eq. (6.11). This is sometimes abbreviated to the following statement: S_n is distributed as $b(n, p)$.

Example 6.4

Random Sampling with Replacement Consider an urn in which there are w white and b black balls of the same size. $\mathcal{E}$ consists of drawing a ball, noting its color, and returning it to the urn. If $\mathcal{E}^n$ is performed by mixing up the balls after each trial, we might well suppose that with these precautions, we have a Bernoulli scheme with parameters n and $p = w/(w + b)$. This scheme is applied to public opinion sampling as follows: people are assigned numbers, and balls marked with these numbers are placed in an urn, sampled with replacement and mixing a total of n times. When a number is drawn, the person so numbered is asked whether he favors some policy (he represents the draw of a white ball) or does not favor the policy (he represents the draw of a black ball). The ratio $w/(w + b)$ is the (typically unknown) proportion of people favoring the policy. In n trials, we will observe r favorable answers for some $0 \le r \le n$, and this is a value of a binomial random variable with parameters n and $w/(w + b)$. On the basis of this sampling, it is common to "estimate" $w/(w + b)$ by the ratio r/n.

EXERCISES 6.3

1. Describe three realistic situations in which one might use the binomial distribution as an appropriate model for the behavior of a random variable.
2. In 30 independent tosses of a balanced coin, what is the probability of getting 25 heads? What is the probability of getting at most 25 heads? (Write an expression for these probabilities. Do not compute.)

3. In a true-false test, a student optimistically adopts a policy of guessing the answer to each question. If we suppose these to be independent trials each with probability of success $\frac{1}{2}$ and if the test has eight questions, what is the probability that the student will score 75% or better?

4. Find the probability that in a family with five children girls will outnumber boys (assume births are independent trials each with probability of boy $\frac{1}{2}$). What would your answer be if you considered a family with seven children? What would your answer be if you considered a family with eight children? Explain your answers.

5. A marksman shoots at a target ten times. He has a probability of $\frac{1}{4}$ of hitting the target each time he fires. What is the probability that he will hit the target at least three times if these trials are taken to be independent?

6. In rolling a balanced die five times, what is the probability of getting at least two sixes if we suppose independent trials?

7. A smooth-talking young man has a probability of $\frac{1}{3}$ of talking a policeman out of giving him a speeding ticket when stopped. The probability that he is stopped during a given weekend is $\frac{1}{2}$. Find the probability that he will receive no tickets at all over five weekends. What is the probability that he will receive at most two tickets over this period? (Assume whatever independence you need.)

8. A factory manufactures a certain kind of bearing. In the past it has been found that on the average 5% of the bearings produced are defective. Find an expression for the probability that in 500 independent inspections of bearings more than 35 bearings are defective. Would it be practicable to compute such an expression?

9. What is the probability that if five people are chosen independently all have their birthdays in the same month? Also find the probability that no two of them are born in the same month and the probability that they are born in exactly three different months. (Assume that all months are equally likely.)

10. If each of two people tosses a balanced coin n times, find the probability that they both get the same number of heads. (Assume whatever independence you need.)

11. Tom and Dick play heads or tails; Tom loses a penny to Dick each time heads appear, and Dick loses a penny to Tom each time tails appear. If they toss the coin four times and these are taken to be independent trials, what is the probability that Tom will have won some money from Dick? What is this probability if they toss the coin seven times? (Assume that the coin is balanced.)

12. A balanced coin is tossed six times. Find the probability that there are more heads among the second, fourth, and sixth tosses than there are among the others if the trials are independent.

13. Suppose we make n independent trials of the following sort: a digit is chosen from among 0, 1, 2, . . . , 9 so that the outcomes are equally likely. What is the probability that no 3 appears? How large must n be so that the probability of the appearance of at least one 3 is more than $\frac{4}{5}$?

14. A balanced coin is tossed eight times. Find the probability of the event "There are at least two heads among the even numbered tosses *and* there are at least two tails in the eight tosses" if tosses are independent.

15. In three consecutive hands, a bridge player had neither an ace nor a king. Assuming independence and that all hands are equally likely, what is the probability of this event? On the basis of your answer, would you say that the player was unlucky?

16. Show that if $p = q = \frac{1}{2}$, the binomial distribution is symmetric, that is to say, the probability of getting exactly r successes is the same as the probability of getting exactly r failures.

17. Let X be a random variable whose distribution is binomial with parameters n and p. Find the most probable value of X. [HINT: We want to find r for which $f_X(r)$ is maximum. To do this, compute the ratio $f_X(r+1)/f_X(r)$, and note the values of r for which this ratio is greater than or equal to one.]

18. Let Ω^4 be sample space for four Bernoulli trials with probability p of success in a given trial. Let X_1 be the number of successes in the first trial (thus, X_1 can assume the values 1 or 0). Similarly, let X_2 be the number of successes in the second trial. Show that X_1 and X_2 are independent random variables. If X_3 and X_4 are analogously defined, what could you say about X_1, X_2, X_3, and X_4?

19. Suppose we make $n + m$ Bernoulli trials with a common probability of success p. Let X be the number of successes in the first n trials, and let Y be the number of successes in the last m trials.
(**a**) What is the probability distribution of X?
(**b**) What is the probability distribution of Y?
(**c**) What is the probability distribution of $X + Y$?
(**d**) Are X and Y independent? Why?
(**e**) Using your answers to (a) through (d), or otherwise, show that

$$\sum_{k=0}^{n} b(k; n, p)b(r-k; m, p) = b(r, n+m, p)$$

where $b(k; n, p)$ denotes the probability that a $b(n, p)$ random variable takes the value k. [HINT: The right side represents the probability that the total number of successes is r. Consider what values X and Y can take if their sum is to be r. Then use (a) through (d).]

***20.** Let $B(k; n, p) = \sum_{r=0}^{k} b(r; n, p)$, with $b(r; n, p)$ as defined in Exercise 19. Thus, $B(k; n, p)$ is the probability of at most k successes in n Bernoulli trials with a common probability p of success in a given trial. Show that

$$B(k; n+1, p) = B(k; n, p) - pb(k; n, p)$$
$$B(k+1; n+1, p) = B(k; n, p) + qb(k+1; n, p)$$

***21.** Show that

$$B(k; n, p) = (n-k)\binom{n}{k}\int_0^q t^{n-k-1}(1-t)^k\,dt$$

(HINT: Differentiate both sides with respect to p, noting that both sides are differentiable functions of p, and show that they have the same derivative. Then use the fundamental theorem of the calculus.)

6.4 Independent trials and the multinomial distribution

In this section, we will consider n independent trials of an experiment $\mathcal{E}$ that may have more than two outcomes. The example of the roulette wheel discussed previously is of this type. Once again, our intention is to study the product probability measure P^n provided by Theorem 6.1.

For simplicity, let us take $\Omega = \{1, 2, \ldots, N\}$ as a sample space for $\mathcal{E}$, and let us suppose that P is a probability measure on Ω determined by $P(\{j\}) = p_j$, $j = 1, 2, \ldots, N$. Of course, we must have $\sum_{j=1}^{N} p_j = 1$. In this case, the probability measure P^n on the sample space Ω^n is called *a multinomial scheme with parameters* $n, p_1, p_2, \ldots, p_N$.

Practical applications of multinomial schemes abound. For example, a die is rolled n times; a bridge hand is drawn and the honor count is tabulated for a total of n times; n people are asked which of five candidates they prefer; and so on. Note that we are considering *independent* trials of an experiment, and any applications must take cognizance of this fact.

We assume that the trials are independent and study P^n on the sample space Ω^n. A sample point $\omega = (\omega_1, \omega_2, \ldots, \omega_n) \in \Omega^n$ is an n-tuple with each entry one of the integers $1, 2, \ldots, N$. By the definition of P^n, we have for $\omega = (\omega_1, \omega_2, \ldots, \omega_n)$,

$$P^n(\{\omega\}) = P(\{\omega_1\})P(\{\omega_2\})\cdots P(\{\omega_n\}) \tag{6.13}$$

Now each ω_i is one of $1, 2, \ldots, N$ so that each factor on the right-hand side of Eq. (6.13) is one of $p_1, p_2, \ldots, p_N$. The probability of $\{\omega\}$ then depends exactly on how many of these factors are p_1, how many are $p_2, \ldots,$ how many are p_N. In particular,

$$P^n(\{\omega\}) = p_1^{n_1}p_2^{n_2}\cdots p_N^{n_N} \tag{6.14}$$

if ω has n_1 entries of 1, n_2 entries of 2, . . . , n_N entries of N, where

$$n_1 + n_2 + \cdots + n_N = n$$

On Ω^n there are N random variables of obvious interest. They are defined by

$$X_j(\omega) \text{ is the number of entries in } \omega \text{ that equal } j;\ j = 1, 2, \ldots, N \tag{6-15}$$

Interpret $X_1, X_2, \ldots, X_N$ in the context of each of the three examples of multinomial schemes given above.

Let us compute here the joint probability distribution of $X_1, X_2, \ldots, X_N$. To begin with, we observe that each of these random variables takes the values $0, 1, 2, \ldots, n$. Taken together, however, it is always true that $X_1(\omega) + X_2(\omega) + \cdots + X_N(\omega) = n$. If $n_1, n_2, \ldots, n_N$ are nonnegative integers for which $n_1 + n_2 + \cdots + n_N = n$, then it is possible that X_1 takes the value n_1, X_2 takes the value n_2, . . . , X_N takes the value n_N. Let

$$A_{n_1,n_2,\ldots,n_N} = \{\omega \mid X_1(\omega) = n_1, X_2(\omega) = n_2, \ldots, X_N(\omega) = n_N\} \tag{6.16}$$

In order to find the joint probability distribution of $X_1, X_2, \ldots, X_N$, we must compute $P^n(A_{n_1,n_2,\ldots,n_N})$ for all choices of $n_1, n_2, \ldots, n_N$ with $n_1 + n_2 + \cdots + n_N = n$. Now

$$P^n(A_{n_1,n_2,\ldots,n_N}) = \sum_{\omega \in A_{n_1,n_2,\ldots,n_N}} P^n(\{\omega\}) \tag{6.17}$$

Any sample point $\omega \in A_{n_1,n_2,\ldots,n_N}$ has exactly n_1 entries of 1, n_2 entries of 2, . . . , n_N entries of N, and, as we have seen in Eq. (6.14), $P^n(\{\omega\})$ for such a sample point is $p_1^{n_1}p_2^{n_2}\cdots p_N^{n_N}$. Accordingly, if $N(A_{n_1,n_2,\ldots,n_N})$ is the number of sample points in $A_{n_1,n_2,\ldots,n_N}$, then

$$P^n(A_{n_1,n_2,\ldots,n_N}) = N(A_{n_1,n_2,\ldots,n_N})p_1^{n_1}p_n^{n_2}\cdots p_N^{n_N} \tag{6.18}$$

The determination of $N(A_{n_1,n_2,\ldots,n_N})$ is a counting problem. Simply, we must determine how many n-tuples ω may be formed having n_1 entries of 1, n_2 entries of 2, . . . , n_N entries of N. This number is equivalent to the number of ways n_1 positions may be chosen for the entries 1, n_2 positions may be chosen for the entries 2, . . . , n_N positions may be chosen for the entries N, viz., $\binom{n}{n_1, n_2, \ldots, n_{N-1}}$. Finally then,

$$P^n(A_{n_1,n_2,\ldots,n_N}) = \binom{n}{n_1, n_2, \ldots, n_{N-1}} p_1^{n_1}p_2^{n_2}\cdots p_N^{n_N} \tag{6.19}$$

In the terminology of the joint probability distribution of $X_1, X_2, \ldots, X_N$, we have found that

$$f_{X_1,X_2,\ldots,X_N}(n_1, n_2, \ldots, n_N) = \binom{n}{n_1, n_2, \ldots, n_{N-1}} p_1^{n_1}p_2^{n_2}\cdots p_N^{n_N} \tag{6.20}$$

for all nonnegative integers $n_1, n_2, \ldots, n_N$ for which

$$n_1 + n_2 + \cdots + n_N = n$$

This joint probability distribution is called the *multinomial distribution,* because the right-hand side of Eq. (6.20) is the general term in the multinomial expansion of $(p_1 + p_2 + \cdots + p_N)^n$.

As we know, it would be possible to find the probability distribution of one of these random variables, say X_1, from the joint probability distribution given by Eq. (6.20). However, it is much easier to find f_{X_1} directly as follows: we forget to differentiate between the outcomes of $\mathcal{E}$ represented by 2, 3, . . . , N and agree to call the outcome represented by 1 a success and all other outcomes failures. We then seek the probability distribution of the number of successes in n independent trials where the probability of a success on a given trial is p_1 and the probability of failure on a given trial is $1 - p_1 = p_2 + p_3 + \cdots + p_N$. In Section 6.3 we found this probability distribution to be a binomial distribution with parameters n and p_1. In general, X_j is distributed as $b(n, p_j)$, $j = 1, 2, \ldots, N$.

Example 6.5

Random Sampling with Replacement (Continued) Consider an urn in which there are c_1 balls of color 1, c_2 balls of color 2, . . . , c_N balls of color N, all balls being the same size. $\mathcal{E}$ consists in drawing a ball, noting its color, and returning it to the urn. If $\mathcal{E}^n$ is performed by mixing up the balls after each trial, we might well suppose that with these precautions, we have a multinomial scheme with parameters

$$n, \frac{c_1}{c_1 + c_2 + \cdots + c_N}, \frac{c_2}{c_1 + c_2 + \cdots + c_N}, \ldots, \frac{c_N}{c_1 + c_2 + \cdots + c_N}$$

This model can be applied to public opinion sampling just as the model in Example 6.4 was. In fact, this model is more general, since we may allow more than two answers to the pertinent question. We might allow yes, no, or don't know, for example. Or, we may ask a person in a television-watchers poll into which category he falls: (1) watching an ABC television program, (2) watching a CBS television program, (3) watching an NBC television program, (4) watching another television program, or (5) not watching television. In n trials for this last question, we may obtain n_1 answers of category 1 (color 1), n_2 answers of category 2 (color 2), . . . , n_5 answers of category 5 (color 5), $n_1 + n_2 + n_3 + n_4 + n_5 = n$. The numbers n_1, n_2, n_3, n_4, and n_5 are values of the random variables X_1, X_2, X_3, X_4, and X_5. They have a multinomial distribution with parameters

$$n, \frac{c_1}{c_1 + c_2 + \cdots + c_5}, \frac{c_2}{c_1 + c_2 + \cdots + c_5}, \ldots, \frac{c_5}{c_1 + c_2 + \cdots + c_5}$$

where $\frac{c_i}{c_1 + c_2 + \cdots + c_5}$ is the (typically unknown) proportion of people in category i, $i = 1, 2, 3, 4, 5$. On this basis of this sampling, it is common to "estimate"

$$\frac{c_i}{c_1 + c_2 + \cdots + c_5} \quad \text{by} \quad \frac{n_i}{n}, \qquad i = 1, 2, 3, 4, 5$$

EXERCISES 6.4

1. Think of five realistic situations in which the random variables of interest have a multinomial distribution.

2. In eight independent throws of a balanced die, what is the probability that sixes and ones each will occur exactly twice, while each of the other scores will occur exactly once.

3. In rolling a balanced die 24 times (assume independent trials), what is the probability of getting each face exactly four times? Similarly, find the probability of getting each face n times in $6n$ rolls of the die.

4. When a big-petalled yellow chrysanthemum is crossbred with a small-petalled white chrysanthemum, the resulting plant may be big-petalled yellow or small-petalled yellow or big-petalled white or small-petalled white. According to Mendel's law of heredity, the respective proportions in which these four phenomena will occur are 9:3:3:1. Find the probability that of 32 plants produced by such a crossbreeding the number of plants in these four categories are respectively 16, 7, 8, 1. (Assume independent trials.)

5. Consider the random variables X_j defined by Eq. (6.15) for $j = 1, 2, \ldots, N$. Are these random variables mutually independent? Why?

6. What is the probability that in ten independent throws of a balanced die a number bigger than 2 will appear at most three times?

7. Among the residents of a city, 40% own their homes, 45% rent their residences on a semipermanent basis, and 15% occupy transient quarters. What is the probability that in a sample of size 500, sampled independently with replacement, exactly 250 own their homes and exactly 200 rent their residence on a semipermanent basis?

8. For the sampling problem in Exercise 7, what is the probability that in the sample of size 500, no more than 275 will own their homes? Would it be wise to compute this probability exactly?

9. The probability that a doctor can diagnose a certain rare disease correctly is $\frac{1}{5}$. If the disease is diagnosed correctly, the probability that the patient is cured is $\frac{9}{10}$. If it is not diagnosed correctly, the patient still has a probability of $\frac{1}{3}$ of being cured in the course of nature. Of 20 patients who consult the doctor, what is the probability that more than six will be cured? (Assume whatever independence you need.)

10. For Exercise 9, what is the probability that exactly six patients will not be diagnosed correctly but cured, and exactly ten patients will be diagnosed correctly and cured. (HINT: There are four possible outcomes for each patient. Regard the 20 patients as 20 multinominal trials in each of which four outcomes are possible.)

11. Suppose that $\Omega = \{1, 2, 3, 4\}$ and that P is equally likely on Ω. Imagine 15 independent trials of an experiment underlying Ω. Compute the conditional probability that $X_2 = 5$ and $X_3 = 3$ given $X_1 = 2$, where X_1, X_2, and X_3 are defined by Eq. (6.15).

12. In a large city, 40% of the voters favor a proposed reform, 30% are against it, and the rest are apathetic. A television newsman decides to poll 20 passers-by at a street corner. What is the probability that among the 20 people he polls the reformists will have an absolute majority, i.e., that there will be more reformists than the other two kinds combined? (Assume independent trials. Why might this be a poor assumption?)

6.5 Dependent trials and the hypergeometric distribution

In the last three sections, we have been examining n independent trials of some experiment $\mathcal{E}$. We have also alluded to the possibility of a similar treatment of a sequence of different experiments $\mathcal{E}_1, \mathcal{E}_2, \ldots, \mathcal{E}_n$ that are performed independently. In many situations, however, a sequence of contemplated experiments $\mathcal{E}_1, \mathcal{E}_2, \ldots, \mathcal{E}_n$ clearly cannot be viewed as being independent in the necessary respects. For example, we might find that the outcome of $\mathcal{E}_1$ has a direct influence on the probabilities of various events concerned with $\mathcal{E}_2$. Let us then consider here the simplest example of *dependent trials*. Although we formulate our discussion in terms of drawing numbered balls from an urn, you should be aware by now that this seemingly special experiment covers a multitude of cases.

Imagine an urn in which there are N balls of similar size numbered $1, 2, \ldots, N$. And consider the experiment of successively drawing $n \leq N$ balls from the urn, *without replacement*, and noting in order which numbers have been drawn. It will be assumed that at each draw (trial), any ball still in the urn has the same probability of being drawn as any other. *Such an experiment will be called random sampling without replacement.* We proceed to construct an appropriate sample space for this composite experiment together with a probability measure on it that reflects the above assumptions.

Let Ω be the set $\{1, 2, \ldots, N\}$. After each of the n draws of our experiment, we may record the number of the ball drawn. At the conclusion of the experiment, we will have a sequence of n integers $\omega_1, \omega_2, \ldots, \omega_n$ such that each $\omega_i \in \Omega$. Each outcome of the experiment could then be described by an ordered n-tuple of the form $(\omega_1, \omega_2, \ldots, \omega_n)$ where $\omega_i \in \Omega$, $i = 1, 2, \ldots, n$. Because the draws are made without replacement, no two of the entries in such an n-tuple can be the same. We shall denote the set of all such ordered n-tuples by $\Omega^{(n)}$. Thus,

$$\Omega^{(n)} = \{\omega = (\omega_1, \omega_2, \ldots, \omega_n) | \omega_i \in \Omega \text{ for each } i, \omega_i \neq \omega_j \text{ for } i \neq j\} \tag{6.21}$$

The set $\Omega^{(n)}$ is sometimes called the *restricted* Cartesian product of Ω with itself n times. In the language of Chapter Three, an outcome of n draws can be viewed simply as a permutation of the N integers $1, 2, \ldots, N$ taken n at a time. Then we readily see that there are exactly $(N)_n = N!/(N - n)!$ sample points in $\Omega^{(n)}$.

We now construct a probability measure P on $\Omega^{(n)}$ that will reflect the assumption made concerning the separate draws. To define P, it is enough to define $P(\{\omega\})$ for each simple event $\{\omega\}$ of $\Omega^{(n)}$. Therefore, let $\omega = (\omega_1, \omega_2, \ldots, \omega_n)$ be a sample point of $\Omega^{(n)}$. Further, let A_1 be the event of $\Omega^{(n)}$ that the first ball drawn is numbered ω_1, A_2 be the event of $\Omega^{(n)}$ that the second ball drawn is numbered $\omega_2, \ldots, A_n$ be the event of $\Omega^{(n)}$ that the nth ball drawn is numbered ω_n. We have

$$\{\omega\} = A_1 \cap A_2 \cap \cdots \cap A_n \tag{6.22}$$

On the first draw, there are N balls in the urn and each, we have assumed, has probability $1/N$ of being drawn. Since A_1 is the event that the first ball drawn is number ω_1, we must assign it probability $1/N$:

$$P(A_1) = \frac{1}{N} \tag{6.23}$$

Next suppose that the event A_1 has occurred, that is, ω_1 has been drawn on the first draw. Then there are $N - 1$ balls left in the urn, and the probability of any one of these being drawn on the second draw is $1/(N - 1)$. Thus the conditional probability of A_2, given A_1, is $1/(N - 1)$. We conclude that we must ensure

$$\begin{aligned} P(A_1 \cap A_2) &= P(A_1)P_{A_1}(A_2) \\ &= \frac{1}{N}\cdot\frac{1}{N - 1} \end{aligned} \tag{6.24}$$

in defining P.

This reasoning can now be repeated. The conditional probability of drawing ω_3, given that ω_1 and ω_2 have been successively drawn on the first two draws, is $1/(N-2)$. Thus we must have $P_{A_1 \cap A_2}(A_3) = 1/(N-2)$, and so

$$P(A_1 \cap A_2 \cap A_3) = P(A_1 \cap A_2)P_{A_1 \cap A_2}(A_3)$$
$$= \frac{1}{N} \cdot \frac{1}{N-1} \cdot \frac{1}{N-2} \tag{6.25}$$

It is clear that by proceeding in this way we will eventually arrive at

$$P(\{\omega\}) = P(A_1 \cap A_2 \cap \cdots \cap A_n)$$
$$= \frac{1}{N} \cdot \frac{1}{N-1} \cdot \frac{1}{N-2} \cdots \frac{1}{N-n+1} \tag{6.26}$$
$$= \frac{1}{(N)_n}$$

To sum up, we see that if a probability measure P on $\Omega^{(n)}$ is to reflect the assumption that at each draw the balls remaining in the urn have equal probabilities of being drawn, then that probability measure P must assign the probability $1/(N)_n$ to each sample point ω of $\Omega^{(n)}$. Clearly, this assignment does indeed define a probability measure. Since $\Omega^{(n)}$ has $(N)_n$ sample points, this measure is the equally likely measure on $\Omega^{(n)}$. Thus our assumptions lead us to choose this particularly simple measure as the appropriate one.

Next we take as fixed the sample space $\Omega^{(n)}$ and the equally likely probability measure P on $\Omega^{(n)}$. We let k be an integer for which $1 \leq k \leq N$ and define a random variable X on $\Omega^{(n)}$ by

$$\text{X is the number of balls drawn from among those numbered } 1, 2, \ldots, k \tag{6.27}$$

Now we compute the probability distribution of X. Clearly X takes only integer values. In particular, no more than k of the balls numbered 1, 2, . . . , k can be drawn, and no more than n of these can be drawn. Thus, if $X(\omega) = r$ for some $\omega \in \Omega^{(n)}$,

$$r \leq k \quad \text{and} \quad r \leq n \tag{6.28}$$

On the other hand, no more than $N - k$ balls numbered $k + 1$, $k + 2$, . . . , N may be drawn, and no more than n of these may be drawn. Since the number of these drawn plus X must be n, it follows that when $X = r$, then $n - r \leq N - k$ and $n - r \leq n$. Therefore,

$$r \geq 0 \quad \text{and} \quad r > n + k - N \tag{6.29}$$

For any integer r satisfying (6.28) and (6.29), let

$$A_r = \{\omega \mid X(\omega) = r\} \tag{6.30}$$

In order to find the probability distribution of X, we must compute $P(A_r)$. Since P is the equally likely measure, it suffices to find $N(A_r)$. Thus, we ask the following question: how many n-tuples are there in $\Omega^{(n)}$ that have r entries from among $1, 2, \ldots, k$ and $n - r$ entries from among $k + 1$, $k + 2, \ldots, N$? The construction of such n-tuples may be done in three stages: first choose r numbers from among $1, 2, \ldots, k$; this can be done in $\binom{k}{r}$ ways; second, choose $n - r$ numbers from among $k + 1, k + 2, \ldots,$ N; this can be done in $\binom{N-k}{n-r}$ ways; finally, order the n numbers chosen, which can be done in $n!$ ways. According to the multiplication principle, $N(A_r) = \binom{k}{r}\binom{N-k}{n-r} n!$. This means that we have

$$f_X(r) = P(A_r) = \frac{N(A_r)}{(N)_n} = \frac{\binom{k}{r}\binom{N-k}{n-r} n!}{\dfrac{N!}{(N-n)!}} = \frac{\binom{k}{r}\binom{N-k}{n-r}}{\binom{N}{n}} \tag{6.31}$$

for r satisfying (6.28) and (6.29).

The probability distribution of X as given by Eq. (6.31) is called *the hypergeometric distribution with parameters* N, k, *and* n. A random variable having this probability distribution is called *a hypergeometric random variable with parameters* N, k, *and* n.

You may have noticed that, although we have the means to answer questions about the order in which balls are drawn in terms of the probability measure P on $\Omega^{(n)}$, the probability distribution of X did not involve this order at all. In fact, the probability distribution of X depends only on the overall composition of the n draws. It is not uncommon to ignore order in this setting. Consider, for example, interpreting the draws of numbered balls as a deal of a hand of 13 cards from a standard deck of 52 one at a time (as is customary). Many questions that are asked concern the final composition of the hand and *not* the order in which the cards appear. To these questions, we would obtain the same answer in terms of P on $\{1, 2, \ldots, 52\}^{(13)}$ as we would obtain by regarding all $\binom{52}{13}$ *combinations* of cards as equally likely (Can you see why?). In particular, if we consider X to be, say, the number of aces that appear in the hand, we will find in either case that X is a hypergeometric random variable with parameters 52, 4, and 13.

Example 6.6

A Waiting Time Distribution Here is a question which does depend on order. Consider P on $\Omega^{(n)}$ as defined above, and let Y be the number of the draw on which the first one of the balls numbered $1, 2, \ldots, k$ appears. In order for Y to be defined over all of $\Omega^{(n)}$, let us ensure that one of the balls numbered $1, 2, \ldots, k$ does appear. The way to do this is to make enough draws, so let $n > N - k$; in other words, draw more balls than there are balls numbered $k + 1, k + 2, \ldots, N$. The possible values of Y may be seen to be $1, 2, \ldots, N - k + 1$. We will find the probability that Y takes each of these values, hence the probability distribution of Y.

Suppose r is an integer and $1 \leq r \leq N - k + 1$. If

$$A_r = \{\omega \mid Y(\omega) = r\}$$

then we want to find

$$P(A_r) = \frac{N(A_r)}{(N)_n}$$

or simply $N(A_r)$. We must find the number of sample points ω in $\Omega^{(n)}$ that have $r - 1$ initial entries drawn from among $k + 1, k + 2, \ldots, N$ and an rth entry drawn from $1, 2, \ldots, k$. We do not need to place a restriction on the last $n - r$ entries of ω. How many sample points are there of this form?

We may construct the sample points in five stages: the first stage consists of selecting $r - 1$ balls from those numbered $k + 1, k + 2, \ldots, N$ in one of $\binom{N-k}{r-1}$ ways; the second stage consists of selecting a ball from those numbered $1, 2, \ldots, k$ in one of k ways; the third stage consists of selecting $n - r$ balls from those not previously selected in one of $\binom{N-r}{n-r}$ ways; the fourth and fifth stages consist of ordering the first $r - 1$ balls in one of $(r - 1)!$ ways and ordering the last balls in one of $(n - r)!$ ways, respectively. That is, by the multiplication principle,

$$N(A_r) = \binom{N-k}{r-1} k \binom{N-r}{n-r} (r-1)!(n-r)!$$

and finally,

$$\begin{aligned} f_Y(r) = \frac{N(A_r)}{(N)_n} &= \frac{\binom{N-k}{r-1} k \binom{N-r}{n-r} (r-1)!(n-r)!}{N!} (N-r)! \\ &= \frac{(N-k)!k(N-r)!(r-1)!(n-r)!(N-n)!}{N!(r-1)!(N-k-r+1)!(n-r)!(N-n)!} \\ &= \frac{k(N-k)!(N-r)!}{N!(N-k-r+1)!} \qquad 1 \leq r \leq N-k+1 \end{aligned}$$

Note that this probability distribution does not depend on n. (We have however taken $n > N - k$.)

EXERCISES 6.5

1. Describe three realistic situations in which the random variable of interest follows a hypergeometric distribution.

2. Five cards are dealt from a bridge deck; it is assumed that this is random sampling without replacement. Find
 (**a**) The probability that exactly two of the cards are kings.
 (**b**) The probability that at least two of the cards are kings.
 (**c**) The probability that no king is drawn on the first two draws.
 (**d**) The conditional probability that at least one king is drawn given that no king is drawn on the first two draws.
 (**e**) The probability that at least two spades are drawn but no more than three black cards are drawn in all.
 (**f**) The probability that exactly two aces are drawn and that one of these two is the ace of spades.
 (**g**) The probability that the hand consists of a sequence of length five, e.g., 2, 3, 4, 5, 6; 7, 8, 9, 10, J; etc.
 (**h**) The probability that the hand consists of two cards of one denomination and three cards of another denomination, e.g., 8, 8, 9, 9, 9.
 (**i**) The probability of getting a royal flush, i.e., A, K, Q, J, 10 of a single suit.

3. A club has 30 members of which 20 are senior members and ten are junior members. A committee of seven is to be elected by random sampling without replacement. Find the probability that
 (**a**) There will be a majority of junior members on the committee.
 (**b**) There will be no more than two junior members on the committee.
 (**c**) Any two members of the committee will have the same seniority.
 (**d**) The conditional probability that there will be a majority of junior members given that at least two of the members of the committee are senior members.

4. An experimental colony of fish has 70 fish, of which 20 are salmon males, 15 are salmon females, 25 are trout males, and the rest are trout females. A random sample of size 12 is chosen without replacement. Find the probability that
 (**a**) There are more salmon than trout in the sample.
 (**b**) There are exactly seven males in the sample.
 (**c**) Males are the majority in the sample.
 (**d**) There are exactly five salmon and exactly three male fish in the sample.
 (**e**) There are exactly five salmon given that there are exactly three males in the sample.

5. Evaluate the waiting time distribution of Example 6.5 when $k = 1$. (Use $0! = 1$, and then derive the distribution directly as a check.)

6. Explain the phenomenon remarked on at the end of Example 6.5.

7. When a certain screw-making machine is in control, it produces 10% defective screws. A box of 100 screws is to be inspected as follows: 15 screws are chosen randomly without replacement and inspected; if more than three of these screws are defective, all the screws in the box are examined. Find the probability that a box will undergo 100% examination even if the process is in control. Find the probability that the box will escape 100% examination even if the process has gone out of control and is producing 20% defectives.

8. An apple grower claims that in each carton of apples he ships there are twelve large and eight medium-sized apples. Twenty cartons are inspected by a buyer in the following way: six apples from each carton are chosen randomly without replacement, and, if more than three of the six are medium size, the carton is rejected. What is the probability that more than ten of the cartons are rejected, assuming that the grower is telling the truth? (Do not try to evaluate this probability.)

9. Suppose that an urn contains balls numbered $1, 2, \ldots, N$. A sample of size n is drawn randomly without replacement from the urn. What is the probability that r *given* balls (say those numbered $l_1, l_2, \ldots, l_r$) are included in the sample?

10. A supermarket shelf has three brands of canned tuna on it, ten cans of the first brand, five of the second, and three of the third. Suppose six cans are chosen by a customer randomly, without replacement. Find the probability that two cans of each brand will be picked.

11. Describe three realistic situations in which the random variable of interest has a hypergeometric distribution with parameters N, k, and n so large that the evaluation of (6.30) is too cumbersome to perform.

***12.** Suppose X has a hypergeometric distribution with parameters N, k, and n. Show that when n is fixed but N and k are allowed to tend to infinity in such a way that $k/N = p$, then $f_X(r)$ given by (6.30) tends to $\binom{n}{r} p^r(1-p)^{n-r}$. Interpret this as an approximation to the hypergeometric distribution. Would this be of use in the situation described in Exercise 11? How?

***13.** Six hundred men live in a dormitory; 300 are underclassmen, and 300 are upperclassmen. A governing committee of five men is to be drawn randomly without replacement. Find the probability that exactly three committee members are upperclassmen and then approximate this probability by using Exercise 12.

***14.** Consider an urn in which there are N balls of which k_1 are of color 1, k_2 are of color 2, $\ldots$, k_q are of color q, $k_1 + k_2 + \cdots + k_q = N$. n balls are sampled randomly without replacement. Find the probability that exactly r_1 of the balls drawn are of color 1, r_2 are of color 2, $\ldots$, r_q are of color q, $r_1 + r_2 + \cdots + r_q = n$. [This is a generalization of the hypergeometric distribution of (6.30), where $q = 2$.]

CHAPTER SEVEN

Expectation

7.1 Introduction

In studying the probability distribution of a random variable, it is often useful to be able to summarize a given aspect of the distribution by means of a single number that then serves to "measure" that aspect. Such a number is called a *parameter* of the distribution. In this chapter, we study two such parameters of a probability distribution, the *expectation* and the *variance*.

The expectation of a random variable is a type of "average" value of that random variable. It is a value about which the possible values of the random variable are scattered. This parameter is our specific concern throughout Sections 7.2 and 7.3.

The variance of a random variable is a measure of the variability of the possible values of that random variable. Roughly speaking, if a random variable assumes values that differ greatly from one another, it has a large variance. If the differences between possible values of a random variable are not large, then the random variable has a small variance. This parameter is discussed more precisely in Sections 7.4 and 7.6.

In Section 7.5 we find expectations and variances for the important binomial and hypergeometric distributions.

7.2 Expectation

Suppose Ω is a sample space, P is a probability measure on the events of Ω, and X is a random variable on Ω. In Chapter Five, we introduced the probability distribution f_X of X. Here we will single out a number, the expectation of X, which may be thought of as an "average" value of X. It will be seen that this number is determined by f_X so that, as long as two random variables have the same probability distribution, they will also have the same expectation. On the other hand, it will be easy to find two random variables that have *different* probability distributions and the *same* expectation. In this sense, the expectation of a random variable measures just one facet of its probability distribution, which might be called the *location* of the probability distribution.

To illustrate the idea behind expectation, let us imagine that the experiment underlying Ω and P is a lottery and that the random variable X is our monetary return from the lottery. That is, we imagine each sample point $\omega \in \Omega$ to be a particular outcome of a lottery, $P(\{\omega\})$ to be its probability of occurrence, and $X(\omega)$ to be our return (in dollars, say) based on this outcome. The possible amounts we win and the probabilities with which we win them are summarized by the probability distribution f_X of X. It would be useful to know, however, what we might expect to obtain in the way of winnings from one performance of the lottery, i.e., we would like an estimate of the worth of this lottery to us. With an estimate of worth, we might be able to give sensible answers to such questions as, "Should we sell our interest in this lottery for d dollars?"

Here is such an estimate based on a frequency interpretation of P. If the lottery were performed a large number of times under similar conditions, say N times, then we could expect that for each $\omega \in \Omega$, approximately $NP(\{\omega\})$ of these times ω would result and we would win $X(\omega)$ dollars. Thus, in N similar lotteries, we would expect to win a total of approximately $\sum_{\omega \in \Omega} NP(\{\omega\})X(\omega)$ dollars. In this sequence of lotteries, the *average* approximate amount won per lottery is

$$\sum_{\omega \in \Omega} \frac{NP(\{\omega\})X(\omega)}{N} = \sum_{\omega \in \Omega} P(\{\omega\})X(\omega) \text{ dollars}$$

We might take this last number as a measure of worth of a single lottery, and we will call it the expectation of X. These remarks merely motivate the definition of the expectation of X. Of course, so far as the mathematical theory is concerned, the definition is logically independent of the frequency interpretation (or any other interpretation) of probability.

We make the following formal definition.

Definition 7.1 Let X be a random variable on a sample space Ω, and let P be a probability measure on Ω. The expectation of X, denoted by $E(X)$, is the real number given by

$$E(X) = \sum_{\omega \in \Omega} X(\omega)P(\{\omega\}) \tag{7.1}$$

It is not uncommon to find $E(X)$ referred to also as the *mean* of the random variable X, but we shall not do so here.

Example 7.1

In the game of craps, two dice are thrown and the scores are noted. One option is to make a field bet on the roll of the dice, say \$1. This bet is lost if a total score of 5, 6, 7, or 8 appears; a win of \$1 is registered if a total score of 3, 4, 9, 10, or 11 appears; a win of \$2 is registered if a total score of 2 appears; and a win of \$3 is registered if a total score of 12 appears. What is the expectation of the return on this bet?

The answer depends, of course, on the probabilities of the various outcomes. If we suppose the roll of the two dice to be independent trials of the roll of a balanced die, the answer can be found as follows. The return depends only on the total score observed. Therefore, we might take a sample space $\Omega = \{2, 3, \ldots, 12\}$ with a probability measure P defined on Ω by $P(\{2\}) = P(\{12\}) = \frac{1}{36}$, $P(\{3\}) = P(\{11\}) = \frac{2}{36}$, $P(\{4\}) = P(\{10\}) = \frac{3}{36}$, $P(\{5\}) = P(\{9\}) = \frac{4}{36}$, $P(\{6\}) = P(\{8\}) = \frac{5}{36}$ and $P(\{7\}) = \frac{6}{36}$. The return X is now defined on Ω by $X(5) = X(6) = X(7) = X(8) = -1$, $X(3) = X(4) = X(9) = X(10) = X(11) = 1$, $X(2) = 2$, and $X(12) = 3$. According to Definition 7.1,

$$\begin{aligned} E(X) &= \sum_{j=2}^{12} X(j)P(\{j\}) \\ &= 2\cdot\tfrac{1}{36} + 1\cdot\tfrac{2}{36} + 1\cdot\tfrac{3}{36} - 1\cdot\tfrac{4}{36} - 1\cdot\tfrac{5}{36} - 1\cdot\tfrac{6}{36} - 1\cdot\tfrac{5}{36} \\ &\qquad + 1\cdot\tfrac{4}{36} + 1\cdot\tfrac{3}{36} + 1\cdot\tfrac{2}{36} + 3\cdot\tfrac{1}{36} = -\tfrac{1}{36} \end{aligned}$$

Although we think of $-\frac{1}{36}$ as an "average" value of X, it is clearly not a possible value of X at all. It is, however, a value in a central location relative to the possible values -1, 1, 2, and 3 of X. In the setting of this bet, it is natural to interpret $-\frac{1}{36}$ from a frequency point of view. Thus, over a long sequence of bets of \$1, say N of them, the bettor can expect a total loss of about $N/36$ dollars and an average loss per bet of about $\frac{1}{36}$ dollars.

Expectation has several useful properties that we will discuss in this section and the next. In particular, we will be interested in finding other ways to compute $E(X)$, since using Eq. (7.1) may be very tedious.

We take as fixed a sample space Ω and a probability measure P on Ω. Any reference to random variables is a reference to random variables on Ω. In the first proposition, we collect together various facts that derive from the definition of expectation and the properties of summation.

Proposition 7.1 (1) $E(X + Y) = E(X) + E(Y)$.
(2) $E(cX) = cE(X)$ for c a real number.
(3) If for each $\omega \in \Omega$, $X(\omega) \geq 0$, then $E(X) \geq 0$.

PROOF: (1)

$$\begin{aligned} E(X + Y) &= \sum_{\omega \in \Omega} (X + Y)(\omega)P(\{\omega\}) \\ &= \sum_{\omega \in \Omega} (X(\omega)P(\{\omega\}) + Y(\omega)P(\{\omega\})) \\ &= \sum_{\omega \in \Omega} X(\omega)P(\{\omega\}) + \sum_{\omega \in \Omega} Y(\omega)P(\{\omega\}) \\ &= E(X) + E(Y) \end{aligned}$$

(2)

$$\begin{aligned} E(cX) &= \sum_{\omega \in \Omega} (cX)(\omega)P(\{\omega\}) \\ &= \sum_{\omega \in \Omega} cX(\omega)P(\{\omega\}) \\ &= c \sum_{\omega \in \Omega} X(\omega)P(\{\omega\}) = cE(X) \end{aligned}$$

(3) If for each $\omega \in \Omega$, $X(\omega) \geq 0$, then for each $\omega \in \Omega$, $X(\omega)P(\{\omega\}) \geq 0$ and $E(X) = \sum_{\omega \in \Omega} X(\omega)P(\{\omega\}) \geq 0$.

Proposition 7.1 is very basic, and we shall make much use of it. Note that (1) may be extended immediately to more than two random variables (see Exercise 11). Therefore,

$$E(X_1 + X_2 + \cdots + X_r) = E(X_1) + E(X_2) + \cdots + E(X_r) \quad (7.2)$$

Equation (7.2) in conjunction with (2) of Proposition 7.1 implies that if $c_1, c_2, \ldots, c_r$ are real numbers, then

$$E(c_1X_1 + c_2X_2 + \cdots + c_rX_r) = c_1E(X_1) + c_2E(X_2) \cdots + c_rE(X_r) \quad (7.3)$$

In principle, $E(X)$ may be computed by using Eq. (7.1). However, if Ω has a large number of sample points, performing the necessary multiplications and additions is quite cumbersome. The next proposition lightens the task.

Proposition 7.2 Let $\mathcal{C} = \{C_1, C_2, \ldots, C_q\}$ be a partition of Ω having the property that for each $\omega \in C_i$, $X(\omega) = z_i$, $i = 1, 2, \ldots, q$. (Thus, X takes the same value z_i at all points in C_i for each i.) Then

$$E(X) = \sum_{i=1}^{q} z_iP(C_i) \quad (7.4)$$

PROOF: According to Eq. (7.1), $E(X) = \sum_{\omega \in \Omega} X(\omega)P(\{\omega\})$. In this sum, we collect together all terms for which $\omega \in C_1$, then all terms for which

$\omega \in C_2, \ldots$, then all terms for which $\omega \in C_q$, and write

$$E(X) = \sum_{\omega \in C_1} X(\omega)P(\{\omega\}) + \sum_{\omega \in C_2} X(\omega)P(\{\omega\}) + \cdots + \sum_{\omega \in C_q} X(\omega)P(\{\omega\}) \tag{7.5}$$

This equation is valid since each $\omega \in \Omega$ is in exactly one of the events $C_1, C_2, \ldots, C_q$. Now, for $\omega \in C_1$, $X(\omega) = z_1$; for $\omega \in C_2$, $X(\omega) = z_2$; . . . ; for $\omega \in C_q$, $X(\omega) = z_q$, so we have

$$E(X) = z_1 \sum_{\omega \in C_1} P(\{\omega\}) + z_2 \sum_{\omega \in C_2} P(\{\omega\}) + \cdots + z_q \sum_{\omega \in C_q} P(\{\omega\}) \tag{7.6}$$

Moreover, $\sum_{\omega \in C_i} P(\{\omega\}) = P(C_i)$ for each one of $i = 1, 2, \ldots, q$. But then Eq. (7.6) reads as

$$E(X) = z_1P(C_1) + z_2P(C_2) + \cdots + z_qP(C_q) \tag{7.7}$$

and this is the required result.

It will be important to us that the numbers $z_1, z_2, \ldots, z_q$ above *need not all be distinct.*

Suppose now that X is a random variable taking the values $x_1, x_2, \ldots, x_k$ on the events $A_1, A_2, \ldots, A_k$, respectively. The partition $\mathcal{A}_X = \{A_1, A_2, \ldots, A_k\}$ of Ω satisfies the condition of Proposition 7.2 for the computation of $E(X)$. In particular, for each $\omega \in A_i$, $X(\omega) = x_i$, $i = 1, 2, \ldots, k$. Applied to the present situation, Eq. (7.4) gives us

$$E(X) = \sum_{i=1}^{k} x_iP(A_i) = \sum_{i=1}^{k} x_if_X(x_i) \tag{7.8}$$

in which we use the fact that $P(A_i) = f_X(x_i)$. Verbally, Eq. (7.8) may be described as follows: the expectation of X equals the sum of the products of the values assumed by X and the respective probabilities with which these values are assumed.

Equation (7.8) is very important. First, it gives $E(X)$ directly in terms of f_X, which makes it clear that if two random variables have the same probability distribution, then they have the same expectation. Second, in many cases, computing $E(X)$ by using Eq. (7.8) is much simpler than computing it by using Eq. (7.1). Indeed, for these reasons, Eq. (7.8) is frequently given as the definition of $E(X)$.

The simplest, and in a sense the most trivial, kind of random variable is one that assumes the *same* value, say c, at *all* points $\omega \in \Omega$. It is customary to call such a random variable a *constant* and to denote it by c. Thus, for example, the symbol 2 stands not only for the integer 2 but also for the random variable whose value at *each* point of Ω is 2, i.e., one that constantly equals 2; the intended meaning is usually made clear by the context. It is trivial to compute the expectation of such a random variable c. Indeed, since it takes only one value c, with probability 1, it follows from Eq. (7.8) that its expectation is c. We write this as

$$E(c) = c \tag{7.9}$$

In the context of a lottery, Eq. (7.9) may be interpreted as follows: if our return from the lottery is constantly equal to c (i.e., if we win a prize of c dollars no matter what the outcome of the lottery), then the expectation of the return is also c. The notion of expectation would have been very unsatisfactory indeed if it did not possess this obvious property.

Another simple type of random variable is an *indicator random variable*. Let A be an event in Ω, and let I_A be its indicator. I_A takes the values 1 and 0 with probabilities $P(A)$ and $P(A')$, respectively. Thus, by Eq. (7.8), we get

$$\begin{aligned} E(I_A) &= 1 \cdot P(A) + 0 \cdot P(A') \\ &= P(A) \end{aligned} \tag{7.10}$$

Note that $E(1) = 1$ is a special case of Eq. (7.10).

Example 7.2

For the matching example of Chapter Three (Example 3.10), there is a random variable that is described as the number of matches. Thus, if two decks of n cards, each numbered $1, 2, \ldots, n$, are played simultaneously one at a time after shuffling, we let X_n be the number of matches observed. X_n clearly takes the values $0, 1, 2, \ldots, n-2, n$ (Why not $n - 1$?), and we have implicitly found the probability distribution of X_n, since we have computed

$$P(\{\omega \mid k \text{ matches occur in } n \text{ plays}\}) = P(\{\omega \mid X_n(\omega) = k\}) = f_{X_n}(k)$$

In particular, if

$$p_0 = 1, \quad p_1 = 0, \quad \text{and} \quad p_n = \frac{1}{2!} - \frac{1}{3!} + \cdots + \frac{(-1)^n}{n!}$$

for $n \geq 2$, then $f_{X_n}(k) = p_{n-k}/k!$ for $k = 0, 1, \ldots, n-2, n$. From Eq. (7.8), it follows that

$$E(X_n) = \sum_{k=0}^{n-2} k \frac{p_{n-k}}{k!} + n \frac{p_0}{n!}$$

Let us compute this sum for small values of n. If $n = 2$, we find that

$$E(X_2) = 2 \cdot \frac{p_0}{2!} = 1$$

If $n = 3$, we find that

$$E(X_3) = 1 \cdot \frac{p_2}{1!} + 3 \cdot \frac{p_0}{3!} = \frac{1}{2} + \frac{1}{2} = 1$$

If $n = 4$, we find that

$$E(X_4) = 1 \cdot \frac{p_3}{1!} + 2 \cdot \frac{p_2}{2!} + 4 \cdot \frac{p_0}{4!}$$

$$= \left(\frac{1}{2} - \frac{1}{3!}\right) + \frac{1}{2} + \frac{1}{6} = 1$$

Thus, in these three cases the expectation of the number of matches is 1. In fact, $E(X_n) = 1$ for every n. This could be shown directly by finding a way to perform the necessary summation above. Rather than do this, we remark that it is easy to show $E(X_n) = 1$ by an indirect method (see Exercise 14).

Example 7.3

It is very useful to be able to compute expectations without going through tedious calculations. In certain cases, the tediousness can be avoided by some simple tricks. As an illustration, let us consider the experiment of drawing a bridge hand of 13 cards from a standard deck of 52, and let us suppose all hands to be equally likely. The honor count of such a hand, say X, is determined by counting 4 points for each ace, 3 points for each king, 2 points for each queen, and 1 point for each jack. Very little thought about the situation will effectively rule out finding $E(X)$ by first finding f_X and then using Eq. (7.8). (Think, for example, about the problem of computing the probability that the honor count is 13. This problem is straightforward if the event that the honor count is 13 is broken down into mutually exclusive events: the hand contains three aces and a jack; the hand contains an ace and three kings; the hand contains a king, four queens, and two jacks; etc. However, finding the relevant probability by adding these mutually exclusive events is very tedious.) Certainly we do not wish to use Eq. (7.1) either. Here is a fast computation of $E(X)$ that we will not spend the time to justify. Imagine a different experiment in which 13 cards are dealt to each of four players and in which all $\binom{52}{13,\ 13,\ 13}$ deals are considered equally likely. Let X_1, X_2, X_3, and X_4 be the honor counts of the four hands. Since the total honor count is 40, it must be that $X_1(\omega) + X_2(\omega) + X_3(\omega) + X_4(\omega) = 40$ for each sample point. According to Eq. (7.2), $E(X_1 + X_2 + X_3 + X_4) = E(X_1) + E(X_2) + E(X_3) + E(X_4)$, and the left side of this equality is clearly 40. It would appear that, with some justification, we could arrive at the fact that $E(X_1) = E(X_2) = E(X_3) = E(X_4)$. If so, each of these must be 10, and we would have found the expectation $E(X)$ of the honor count in a single hand to be 10.

Alternatively, we might find $E(X)$ as follows. Let I_1, I_2, I_3, and I_4 be the indicator random variables of the events that the hand contains the ace of spades, the ace of hearts, the ace of diamonds, and the ace of clubs, respectively. Similarly, let I_5, I_6, I_7, and I_8 be the corresponding indicator random variables concerned with the kings in each suit, I_9, I_{10}, I_{11}, and I_{12} the queens, and I_{13}, I_{14}, I_{15}, and I_{16}, the jacks. Then

$$X = 4 \sum_{j=1}^{4} I_j + 3 \sum_{j=5}^{8} I_j + 2 \sum_{j=9}^{12} I_j + \sum_{j=13}^{16} I_j$$

(Check this for each ω.) Using Eq. (7.3), we find

$$E(X) = 4 \sum_{j=1}^{4} E(I_j) + 3 \sum_{j=5}^{8} E(I_j) + 2 \sum_{j=9}^{12} E(I_j) + \sum_{j=13}^{16} E(I_j)$$

Now the expectation of each I_j is, by Eq. (7.9), precisely the probability that the hand contains a specified card, viz., $\frac{1}{4}$. But then

$$E(X) = 4 + 3 + 2 + 1 = 10$$

In conclusion, we remark that $E(X)$ is often called a measure of location. It is clear that $E(X)$ falls between the smallest value, say x_1, assumed by X and the largest value, say x_k, assumed by X. Thus, a knowledge of $E(X)$ gives us a rough idea of the size of the possible values of X. If we imagine plotting these values, say $x_1, \ldots, x_k$, on a straight line, a knowledge of $E(X)$ would determine roughly where these numbers $x_1, \ldots, x_k$ would be located with respect to the origin. As a parameter of location, $E(X)$ has natural properties. For example, if instead of considering X we consider the random variable $X + c$, whose values are $x_1 + c, x_2 + c, \ldots, x_k + c$, it is clear that these values are obtained just by displacing the values $x_1, \ldots, x_k$ by a constant amount c. The property $E(X + c) = E(X) + c$ says that the value of $E(X)$ also gets "displaced" by c. This of course is how we would expect a parameter of location to behave. Again, if instead of X we consider the random variable cX, where c is a constant, we see that the possible values of cX are $cx_1, cx_2, \ldots, cx_k$. If we imagine plotting the values of cX on a straight line, it is clear that they are the same as the values of X, scaled by a factor c. We would then expect that a parameter of location would also get scaled by the same factor. This expectation is borne out by the property $E(cX) = cE(X)$ of the expectation. Further, we might consider an analogy between the concept of expectation and the concept of center of gravity in mechanics. If we imagine masses of size $f_X(x_i)$ being placed at points x_i on the line, $i = 1, 2, \ldots, k$, then $E(X)$ is exactly what physicists call the center of gravity of this mass distribution. The equation $E(X + c) = E(X) + c$ is the analogue of the fact known to physicists that if the location of all the masses of the mass distribution is displaced through a distance c, then their center of gravity also is displaced through a distance c.

EXERCISES 7.2

1. Give examples of two random variables having different probability distributions and the same expectation.

2. For two independent throws of a balanced die, find the expectations of the following random variables.
   (a) The score on the first throw.
   (b) The total score.
   (c) The larger of the two scores.
   (d) The smaller of the two scores.
   (e) The first score minus the second.
   (f) The larger score minus the smaller.
   (g) Twice the larger score minus the second.
   (h) The total number of sixes that appear.
   (i) The total number of fives and sixes that appear.

3. Consider a random sample of size 13 drawn without replacement from a deck of 52 cards. Find the expectations of the following random variables.
   (a) The number of aces drawn.
   (b) The number of kings drawn.
   (c) The number of aces and kings drawn.
   (d) The number of cards drawn.
   (e) The number of hearts drawn.
   (f) The number of red cards drawn.
   (g) The number of heart aces drawn.
   (h) The number of red cards drawn minus the number of black cards drawn.
   (i) The number of red cards drawn minus the number of hearts drawn.
   (Use indicator random variables as in Example 7.3.)

4. For four independent tosses of a balanced coin, find the expectation of the number of heads that appear. Also find the expectation of the number of heads minus the number of tails.

5. Suppose an experiment $\mathcal{E}$ produces a success with probability $\frac{1}{4}$. What is the expected number of successes in four independent trials of $\mathcal{E}$.

6. One card is drawn at random from a standard 52-card deck. If the card drawn is the seven of diamonds we win $420; if it is any other card we win nothing. On the basis of expectation, should we be willing to sell our interest in this experiment for $5? For $10?

7. A Good Humor man finds that the demand for ice cream bars is 100 with probability $\frac{2}{3}$ (the day is warm) and 35 with probability $\frac{1}{3}$ (the day is cold).

If he pays \$.05 per bar, sells bars at \$.10, and must discard unsold bars at the end of the day, what is his expected profit (or loss) if he orders 35 bars? 50 bars? 100 bars?

8. A game played by two persons is said to be "fair" if the expectation of return is zero for each player. Tom and Dick toss a balanced die, and Tom agrees to pay Dick \$6 if the score registered is less than or equal to 2. How much should Dick pay Tom when the score is larger than 2 in order that the game be fair?

9. A lottery has three prizes of \$1000, \$500, and \$100. Five thousand tickets are to be sold and the winning tickets will be drawn as a random sample of size three from them.

(**a**) What price should be charged per ticket if this amount is to be the same as the expectation of return on a single ticket?

(**b**) Suppose a person bought his ticket at the price determined in (**a**) and then learned that a total of 4000 tickets were sold. Should he complain?

(**c**) If 5000 tickets are sold at the price determined in (**a**), how much profit will the sponsors of the lottery make? Why?

(**d**) A man is approached about buying the last of 5000 tickets at the price determined in (**a**). Is he justified in thinking that his expectation of return is less than the price of the ticket since all winning tickets may have already been sold?

10. Let P be a probability measure on a sample space Ω, and let X be a random variable on Ω. Show that if for each $\omega \in \Omega$, $X(\omega) \geq c$, then $E(X) \geq c$. Show that if for each $\omega \in \Omega$, $X(\omega) \leq d$, then $E(X) \leq d$.

11. Show that Eq. (7.2) holds. Use (1) of Proposition 7.1 and induction.

12. Suppose one bus arrives at a given corner each hour. With probability $\frac{1}{6}$, it arrives at 10 minutes past the hour, with probability $\frac{1}{2}$ it arrives at 25 minutes past the hour, and with probability $\frac{1}{3}$ it arrives at 50 minutes past the hour. What is the expectation of the waiting time for a man who arrives at the corner on the hour? At 15 minutes past the hour? Can you find the best time for the man to arrive at the corner (best in the sense that it minimizes the expectation of the waiting time)?

13. For the waiting time distribution in Example 6.6, calculate the expectation of waiting time for N draws without replacement from N balls with $k = 1$. Repeat for N draws without replacement from N balls with $k = 2$.

14. Show that in Example 7.2, $E(X_n) = 1$ for all n. (Let I_j be the indicator random variable of the event that there is a match on the jth play, $j = 1, 2, \ldots, n$, and note that $X_n = \sum_{j=1}^{n} I_j$.)

15. Suppose a random variable X on Ω has the property that for each $\omega \in \Omega$, $X(\omega) \geq 0$. If P is a probability measure on Ω for which $E(X) = 0$, show that $P(\{\omega | X(\omega) > 0\}) = 0$.

16. Let P be a probability measure on a sample space Ω, and let X be a random variable on Ω. The expectation of X calculated under the probability measure P_A for an event A from Ω is called the *conditional expectation* of X given A. If this is denoted by $E_A(X)$, then $E_A(X) = \sum_{\omega \in \Omega} X(\omega)P_A(\{\omega\})$. Find the conditional expectations of the random variables in Exercise 2 given that the first score is 4. Do the same thing given that the two scores are the same.

17. Let P be a probability measure on a sample space Ω, and let X and Y be two random variables on Ω. Suppose X takes the values $x_1, x_2, \ldots, x_k$ on the events $A_1, A_2, \ldots, A_k$, respectively. The function defined on $x_1, x_2, \ldots, x_k$ with value $E_{A_i}(Y)$ at x_i, $i = 1, 2, \ldots, k$, is denoted by $E_X(Y)$ and is called the *regression function* of Y on X. Show that $\sum_{i=1}^{k} E_X(Y)(x_i)f_X(x_i) = E(Y)$.

7.3 Expectations and moments

In this section, we find some further reductions for the problem of computing expectations. At the same time, we give special attention to expectations of some specific random variables.

Suppose again that Ω is a fixed sample space and that P is a probability measure on Ω. X and Y will be two random variables on Ω, X taking the values $x_1, x_2, \ldots, x_k$ on the events $A_1, A_2, \ldots, A_k$, respectively, and Y taking the values $y_1, y_2, \ldots, y_m$ on the events $B_1, B_2, \ldots, B_m$, respectively. Thus, $\mathcal{A}_X = \{A_1, A_2, \ldots, A_k\}$ and $\mathcal{A}_Y = \{B_1, B_2, \ldots, B_m\}$ are the partitions of Ω induced by X and Y. We write, as usual, f_X, f_Y, and $f_{X,Y}$, respectively, for the probability distribution of X, of Y, and for the joint probability distribution of X and Y.

To begin with, consider a random variable of the form $g(X)$. How should we compute $E(g(X))$? If we know the probability distribution f_X of X, then in principle it is possible to find $f_{g(X)}$ from f_X (see Exercise 13 of Section 5.4). Accordingly, we could do this and then apply Eq. (7.8). Fortunately, this process can be avoided, and $E(g(X))$ may be found directly from f_X as is evident from the next proposition.

Proposition 7.3 For any random variable of the form $g(X)$,

$$E(g(X)) = \sum_{i=1} g(x_i)f_X(x_i) \tag{7.11}$$

PROOF: We appeal to Proposition 7.2. For each $\omega \in A_i$, $X(\omega) = x_i$, $i = 1, 2, \ldots, k$. Therefore, for each $\omega \in A_i$, $(g(X))(\omega) = g(X(\omega)) = g(x_i)$, $i = 1, 2, \ldots, k$. Thus, the random variable $g(X)$ takes the same value $g(x_i)$ at each point $\omega \in A_i$, $i = 1, 2, \ldots, k$. Applying Proposition 7.2 to the random variable $g(X)$ and the partition $\mathcal{A}_X$, we find

$$E(g(X)) = \sum_{i=1}^{k} g(x_i)P(A_i) = \sum_{i=1}^{k} g(x_i)f_X(x_i) \tag{7.12}$$

The probability distribution of X thus determines not only $E(X)$ but also the expectation of any function $g(X)$ of X. If X and Y have the same probability distribution, it follows that $g(X)$ and $g(Y)$ have the same expectation for any function g.

The above discussion applies in particular to $g(X) = X^n$ for any choice of a positive integer n. The expectations $E(X^n)$, $n = 1, 2, \ldots$, play an important role in probability theory and in statistics.

Definition 7.2 For each positive integer n, the number $E(X^n)$ is called the nth moment of (the probability distribution of) X about the origin and is denoted by μ_n. Thus,

$$\mu_n = E(X^n) = \sum_{i=1}^{k} (x_i)^n f_X(x_i) \tag{7.13}$$

Of course, if $n = 1$, we see that $\mu_1 = E(X)$. Since this first moment is the one most frequently encountered, the subscript is often dropped and we write $\mu = E(X)$. The moments μ_n are of some theoretical importance. For example, it may be shown that if X and Y have the same moments $[E(X^n) = E(Y^n),\ n = 1, 2, \ldots]$ then X and Y have the same probability distribution (we already know that the converse is true). The proof of this is sketched in Exercise 11. In practice, however, moments of higher order than four are rarely used; even the third and the fourth moments are used infrequently in comparison with the first two.

Example 7.4

Here is a simple probability distribution for which all moments are easily found. Let $f_X(0) = \frac{1}{4} = f_X(2)$, and let $f_X(1) = \frac{1}{2}$ (for example, the random variable X might be the number of heads appearing in two independent tosses of a balanced coin). Now

$$E(X^n) = 1^n\tfrac{1}{2} + 2^n\tfrac{1}{4} = \tfrac{1}{2}(1 + 2^{n-1})$$

Thus, $\mu_n = \frac{1}{2}(1 + 2^{n-1})$ for $n = 1, 2, \ldots$. You might at this point reflect on the difficulties inherent in finding all moments of some general random variables.

The following is an analogue of Proposition 7.3 for random variables that are functions of X and Y.

Proposition 7.4 For any random variable of the form $g(X,Y)$,

$$E(g(X, Y)) = \sum_{i=1}^{k} \sum_{j=1}^{m} g(x_i, y_j) f_{X,Y}(x_i, y_j) \tag{7.14}$$

PROOF: For each $\omega \in A_i \cap B_j$, $X(\omega) = x_i$ and $Y(\omega) = y_j$, $i = 1, 2, \ldots, k$ and $j = 1, 2, \ldots, m$. Then for each $\omega \in A_i \cap B_j$, $(g(X, Y))(\omega) = g(X(\omega), Y(\omega)) = g(x_i, y_j)$, $i = 1, 2, \ldots, k$, $j = 1, 2, \ldots, m$. Applying Proposition 7.2 to the random variable $g(X, Y)$ and the partition $\mathcal{A}_X \wedge \mathcal{A}_Y$, we find

$$\begin{aligned} E(g(X, Y)) &= \sum_{i=1}^{k} \sum_{j=1}^{m} g(x_i, y_j) P(A_i \cap B_j) \\ &= \sum_{i=1}^{k} \sum_{j=1}^{m} g(x_i, y_j) f_{X,Y}(x_i, y_j) \end{aligned} \tag{7.15}$$

In Eq. (7.15) we have identified $P(A_i \cap B_j)$ with the joint probability distribution of X and Y evaluated at (x_i, y_j), and this is the required result.

The special case of Proposition 7.4 that we wish to single out is the case of $g(X, Y) = XY$. Equation (7.14) now becomes

$$E(XY) = \sum_{i=1}^{k} \sum_{j=1}^{m} x_i y_j f_{X,Y}(x_i, y_j) \tag{7.16}$$

This expression for $E(XY)$ simplifies greatly if X and Y are *independent* random variables.

Proposition 7.5 If X and Y are independent random variables, then

$$E(XY) = E(X)E(Y) \tag{7.17}$$

PROOF: By Eq. (7.16), $E(XY) = \sum_{i=1}^{k} \sum_{j=1}^{m} x_i y_j f_{X,Y}(x_i, y_j)$. Now the independence of X and Y means that for each of $i = 1, 2, \ldots, k$ and $j = 1, 2, \ldots, m$,

$$\begin{aligned} f_{X,Y}(x_i, y_j) &= P(A_i \cap B_j) = P(A_i)P(B_j) \\ &= f_X(x_i) f_Y(y_j) \end{aligned} \tag{7.18}$$

Therefore, we obtain

$$\begin{aligned} E(XY) &= \sum_{i=1}^{k} \sum_{j=1}^{m} x_i y_j f_{X,Y}(x_i, y_j) = \sum_{i=1}^{k} \sum_{j=1}^{m} x_i y_j f_X(x_i) f_Y(y_j) \\ &= \Big(\sum_{i=1}^{k} x_i f_X(x_i)\Big)\Big(\sum_{j=1}^{m} y_j f_Y(y_j)\Big) = E(X)E(Y) \end{aligned} \tag{7.19}$$

In words, Proposition 7.5 states that *if* X and Y are independent, *then* the expectation of their product is equal to the product of their expectations. Note particularly that the proposition is *not* necessarily true if X and Y are not independent. In this case, $E(XY)$ would have to be computed from Eq. (7.16).

Also, note that the converse of Proposition 7.5 is *not* true. That is, it is possible to have random variables X and Y such that $E(XY) = E(X)E(Y)$, and such that X and Y are *dependent.* We shall show this by an example below.

Two further results concerning expectation may be seen to hold. The first is the natural extension of Propositions 7.3 and 7.4 to a function $g(X_1, X_2, \ldots, X_r)$ of r random variables (see Exercise 4). The second result, which we do not prove here, is that Proposition 7.5 may be extended to a statement that the expectation of the product of $r \geq 2$ *mutually independent* random variables is the product of their respective expectations. (See Exercise 5.)

Example 7.5

Consider a random variable X whose distribution is symmetric about zero. By this we mean that whenever x is a possible value of X, then so also is $-x$, and further the probabilities with which X assumes these values are equal. In other words, $f_X(x) = f_X(-x)$. Such random variables occur quite often. For example, if X is the number of heads minus the number of tails in three tosses of a fair coin, then the possible values of X are $-3, -1, +1, +3$, and we have $f_X(-3) = f_X(3) = \frac{1}{8}$, $f_X(-1) = f_X(1) = \frac{3}{8}$. Thus, the distribution of X is symmetric about zero. If X is such a random variable, it is easily shown that all the moments of X of odd order are zero. That is, for each $n = 0, 1, 2, \ldots$, we have $E(X^{2n+1}) = 0$. This may be seen in Eq. (7.13), since every term of the form $x^{2n+1}f_X(x)$ there is cancelled by the term $(-x)^{2n+1}f_X(-x) = -x^{2n+1}f_X(-x)$; (the latter term must occur because of the symmetry assumption).

Let us now consider a random variable X with a symmetric distribution, and let $Y = X^2$. Then we note that $E(X) = 0$ and $E(XY) = E(X^3) = 0$. Therefore, $E(XY) = E(X)E(Y)$. However, X and Y are patently dependent random variables, since Y is the square of X. (There is one exception to this statement; see Exercise 8.) This example shows that the converse of Proposition 7.5 does not hold. Of course, in this example, we could have considered instead of the pair of random variables X and X^2, any other pair of the form X^{2q+1} and X^{2r}, where q and r are positive integers.

EXERCISES 7.3

1. Let X be the score registered on one throw of a balanced die. Give the probability distribution f_X of X, and find $E(X)$, $E(2X + 3)$, $E(X^2)$, $E(3X^2 + 2X + 1)$, and $E(1/X)$.

2. Find all the moments of an indicator random variable.

3. Let X be the number of heads and Y the largest number of consecutive heads in five independent tosses of a balanced coin ($Y = 0$ if no heads appear).
(**a**) Find f_X and f_Y.
(**b**) Find $f_{X,Y}$.
(**c**) Find $E(X)$, $E(Y)$, $E(XY)$, and $E(X^2 + Y)$.
Are X and Y independent?

4. Show that if $X_1, X_2, \ldots, X_r$ have a joint probability distribution $f_{X_1, X_2, \ldots X_r}$, then $E(g(X_1, X_2, \ldots, X_r)) = \sum g(x_1, x_2, \ldots, x_r) f_X(x_1, x_2, \ldots, x_r)$ where the sum extends over all r-tuples $(x_1, x_2, \ldots, x_r)$ such that x_1 is a possible value of X_1, x_2 is a possible value of X_2, . . . , x_r is a possible value of X_r.

5. Show that if $X_1, X_2, \ldots, X_r$ are mutually independent random variables, then $E(X_1 \cdot X_2 \cdot \cdots \cdot X_r) = E(X_1)E(X_2) \cdots E(X_r)$.

6. Construct an example in which X and Y are two dependent random variables, $E(XY) = E(X)E(Y)$ and $E(XY) \neq 0$.

7. Is it generally true that $h(E(X)) = E(h(X))$?

8. In Example 7.5, show that X and $Y = X^2$ are independent only in the special case that Y takes exactly one value.

9. Suppose a random variable X has moments $\mu_1, \mu_2, \ldots, \mu_n, \ldots$. Show that cX has nth moment $c^n \mu_n$. Show that $X + c$ has nth moment $\sum_{r=0}^{n} \binom{n}{r} c^r \mu_{n-r}$.

***10.** Suppose that X has moments $\mu_1, \mu_2, \ldots, \mu_n \ldots$, and that Y has moments $\nu_1, \nu_2, \ldots, \nu_n, \ldots$. If X and Y are independent, show that $X + Y$ has nth moment $\sum_{r=0}^{n} \binom{n}{r} \nu_r \mu_{n-r}$.

***11.** Let $x_1, x_2, \ldots, x_k$ be any *distinct* real numbers. Show that for each i, $1 \leq i \leq k$, we can find a polynomial F_i for which $F_i(x_i) = 1$ and $F_i(x_j) = 0$ for $j \neq i$. [HINT: Let $F_i(x)$ be a constant multiple of $(x - x_1)(x - x_2) \cdots (x - x_{i-1})(x - x_{i+1}) \cdots (x - x_k)$.]

***12.** Let X and Y be two random variables on Ω, and let P be a probability measure on Ω. Suppose (i) X and Y assume the same values, and (ii) $E(X^n) = E(Y^n)$ for all positive integers n. Show that $f_X = f_Y$. [HINT: Let $x_1, x_2, \ldots, x_k$ be the values assumed by X and Y, and let $p_j = f_X(x_j)$, $q_j = f_Y(x_j)$, $j = 1, 2, \ldots, k$. Show that (ii) implies that if F is *any* polynomial, $E(F(X)) = E(F(Y))$. In particular, for each i, $E(F_i(X)) = E(F_i(Y))$, where F_i is the polynomial constructed in Exercise 10. Evaluate $E(F_i(X))$ and $E(F_i(Y))$ to obtain the desired conclusion.] Can you prove this result without assumption (i)? In the version above, were the hypotheses used fully? If not, state a weaker form of the assumptions that would suffice to give the same conclusion.

7.4 Variance and covariance

Let P be a probability measure on Ω, and suppose X is a random variable on Ω. The expectation of X has been defined and found to be determined by the probability distribution f_X of X. In a certain sense, $E(X)$ locates f_X. Here we define another quantity depending on f_X which will, in a sense, measure the "spread" of f_X. It is called the variance of (the probability distribution of) X.

Definition 7.3 Let P be a probability measure on a sample space Ω, and let X be a random variable on Ω with expectation $E(X) = \mu$. The *variance* of X, denoted by $V(X)$, is the real number given by

$$V(X) = E((X - \mu)^2) \tag{7.20}$$

Clearly, $(X - \mu)^2(\omega) = (X(\omega) - \mu)^2 \geq 0$ for each ω, so by (3) of Proposition 7.1, $V(X) \geq 0$. The nonnegative square root of $V(X)$ is called the *standard deviation* of X and is denoted by σ_X. Thus,

$$\sigma_X^2 = V(X) \tag{7.21}$$

It is apparent from Eq. (7.20) that the variance of X is simply the expectation of a particular function of X, viz., $(X - \mu)^2$. From Eq. (7.11), $V(X)$ may be computed as

$$V(X) = E((X - \mu)^2) = \sum_{i=1}^{k} (x_i - \mu)^2 f_X(x_i) \tag{7.22}$$

which can be rewritten by expanding each $(x_i - \mu)^2$ as $x_i^2 - 2\mu x_i + \mu^2$ and then adding each portion individually. We find [recall that $\mu = E(X)$]

$$\begin{aligned} V(X) &= \sum (x_i - \mu)^2 f_X(x_i) = \sum (x_i^2 - 2\mu x_i + \mu^2) f_X(x_i) \\ &= \sum x_i^2 f_X(x_i) - 2\mu \sum x_i f_X(x_i) + \mu^2 \sum f_X(x_i) \\ &= E(X^2) - 2\mu E(X) + \mu^2 = E(X^2) - \mu^2 = \mu_2 - \mu^2 \end{aligned} \tag{7.23}$$

That is, Eq. (7.23) gives $V(X)$ as the difference between the second moment of X about the origin and the square of the first moment of X about the origin. It is often convenient to compute $V(X)$ by finding μ and μ_2 first, and then using Eq. (7.23).

Example 7.6

Let us compute $V(X_n)$ for small values of n, where X_n is the number of matches as given in Example 7.2. For $n = 2$, 3, and 4, we found the expectation of X_n to be one. According to Eq. (7.23), $V(X_n)$ could then be found from $E(X_n^2)$, $n = 2, 3, 4$. Now with $n = 2$, $f_{X_2}(0) = \frac{1}{2}$ and $f_{X_2}(2) = \frac{1}{2}$; with $n = 3$, $f_{X_3}(0) = \frac{1}{3}$, $f_{X_3}(1) = \frac{1}{2}$, and $f_{X_3}(3) = \frac{1}{6}$; with $n = 4$, $f_{X_4}(0) = \frac{3}{8}$, $f_{X_4}(1) = \frac{1}{3}$, $f_{X_4}(2) = \frac{1}{4}$, and $f_{X_4}(4) = \frac{1}{24}$. Thus, $E(X_2^2) = 4 \cdot \frac{1}{2} = 2$, $E(X_3^2) = \frac{1}{2} + 9 \cdot \frac{1}{6} = 2$, and $E(X_4^2) = \frac{1}{3} + 4 \cdot \frac{1}{4} + 16 \cdot \frac{1}{24} = 2$. But then $V(X_2) = V(X_3) = V(X_4) = 1$. It is possible to use the general expression for $f_{X_n}(k)$ to show that $V(X_n) = 1$ for all choices of n. However, an indirect method will show that this can be done quite easily (see Exercise 8). Of course, then the standard deviation of each X_n is also one.

We give two elementary properties of variance in the next proposition, where we continue to denote $E(X)$ by μ.

Proposition 7.6 (1) For c a real number, $V(X + c) = V(X)$.
(2) For c a real number, $V(cX) = c^2V(X)$.

PROOF: (1) Let $Y = X + c$. Then $E(Y) = E(X) + c = \mu + c$. $V(Y)$ is given by $E((Y - (\mu + c))^2) = E((X + c - (\mu + c))^2) = E((X - \mu)^2) = V(X)$.

(2) Let $Z = cX$. Then $E(Z) = cE(X) = c\mu$. $V(Z)$ is given by $E((Z - c\mu)^2) = E((cX - c\mu)^2) = E(c^2(X - \mu)^2) = c^2E(X - \mu)^2 = c^2V(X)$.

As a consequence of (1) and (2), $\sigma_{X+c} = \sigma_X$ and $\sigma_{cX} = |c|\sigma_X$.

$V(X)$ is determined by the probability distribution f_X of X. $V(X)$ is commonly said to measure the extent of the "spread" or "dispersion" of the distribution f_X around its expectation. Of course, by definition, $V(X)$ is an average value of $(X - \mu)^2$, and, in general, the larger the possible values of this random variable [i.e., the values $(x_i - \mu)^2$], the larger is $V(X)$. Conversely, a large value for $V(X)$ means that the deviations of X from its expectation μ are frequently quite large, which means that the values of X are more "spread out" or "dispersed." In this sense, $V(X)$ measures the spread of f_X. $V(X)$ does indeed possess some of the properties that we would expect a measure of spread to have. For example, suppose we consider the random variable $Y = X + c$, where c is a constant. Then $E(Y) = E(X) + c$ and so $Y - E(Y) = X - E(X)$. Thus, the deviations of Y from its expectation are the same as those of X from *its* expectation. Therefore, though the locations of the distributions of X and Y differ, we would expect these two distributions to have the same "spread"; in other words, a measure of spread should have the same value in both these cases. The property $V(X + c) = V(X)$ of the variance shows that the variance behaves in the expected manner.

If we imagine masses of size $f_X(x_i)$ placed at points x_i on the axis,

$V(X)$ is what physicists call the moment of inertia of the mass distribution about its center of gravity. The property $V(X + c) = V(X)$ of the variance is then the analogue of the fact known to physicists that, if the locations of all the masses in a mass distribution are displaced through a constant distance c, the moment of inertia of the mass distribution about *its center of gravity* does not change. How would you, in a similar way, interpret the property $V(cX) = c^2V(X)$?

Of course, the variance is not the only possible measure of the dispersion. There are other measures as well. However, they will not concern us. The variance is the one most commonly used and most convenient.

Consider now two random variables X and Y on Ω. We have seen, and have made considerable use of the fact, that $E(X + Y) = E(X) + E(Y)$. It might then be asked if one can make a corresponding statement about $V(X + Y)$. Before answering this question, we make the following definition.

Definition 7.4 Let P be a probability measure on a sample space Ω, and let X and Y be random variables on Ω with expectations μ_X and μ_Y, respectively. The *covariance* of X and Y, denoted by Cov (X, Y), is the real number given by

$$\text{Cov}\,(X, Y) = E((X - \mu_X)(Y - \mu_Y)) \tag{7.24}$$

Clearly, the notion of covariance is symmetric in X and Y so that

$$\text{Cov}\,(X, Y) = \text{Cov}\,(Y, X)$$

You might also notice that

$$\text{Cov}\,(X, X) = V(X)$$

On the basis of Eq. (7.15), Cov (X, Y) may be computed as

$$\begin{aligned} \text{Cov}\,(X, Y) &= E((X - \mu_X)(Y - \mu_Y)) \\ &= \sum_{i=1}^{k} \sum_{j=1}^{m} (x_i - \mu_X)(y_j - \mu_Y) f_{X,Y}(x_i, y_j) \end{aligned} \tag{7.25}$$

Alternatively, we might expand $(X - \mu_X)(Y - \mu_Y)$ and then compute expectations term by term. We then obtain, recalling that $E(X) = \mu_X$ and $E(Y) = \mu_Y$,

$$\begin{aligned} \text{Cov}\,(X, Y) &= E((X - \mu_X)(Y - \mu_Y)) \\ &= E(XY - \mu_X Y - \mu_Y X + \mu_X\mu_Y) \\ &= E(XY) - \mu_X E(Y) - \mu_Y E(X) + \mu_X\mu_Y \\ &= E(XY) - \mu_X\mu_Y \end{aligned} \tag{7.26}$$

Equation (7.23) was obtained in the same way, and it is worth noticing its resemblance to Eq. (7.26). According to Proposition 7.5, if X and Y are *independent*, then $E(XY) = E(X)E(Y) = \mu_X\mu_Y$. Therefore, Eq. (7.26) says that *independent* random variables have covariance zero. However, it is possible to have $\text{Cov}(X, Y) = 0$ when X and Y are *dependent* also (recall Example 7.5).

Now we turn to a computation of $V(X + Y)$.

Proposition 7.7 $V(X + Y) = V(X) + V(Y) + 2\,\text{Cov}(X, Y)$.

PROOF: According to its definition,

$$V(X + Y) = E((X + Y - (\mu_X + \mu_Y))^2)$$

where again, $E(X) = \mu_X$ and $E(Y) = \mu_Y$. Now observe that $V(X + Y)$ may be rewritten as

$$\begin{aligned} V(X + Y) &= E((X + Y - (\mu_X + \mu_Y))^2) \\ &= E((X - \mu_X + Y - \mu_Y)^2) \\ &= E((X - \mu_X)^2 + (Y - \mu_Y)^2 + 2(X - \mu_X)(Y - \mu_Y)) \\ &= E((X - \mu_X)^2) + E((Y - \mu_Y)^2) + 2E((X - \mu_X)(Y - \mu_Y)) \\ &= V(X) + V(Y) + 2\,\text{Cov}(X, Y) \end{aligned} \tag{7.27}$$

As mentioned above, independence of X and Y implies

$$\text{Cov}(X, Y) = 0$$

The following corollary to Proposition 7.7 is then immediate.

Corollary 7.1 If X and Y are independent random variables, then

$$V(X + Y) = V(X) + V(Y) \tag{7.28}$$

In words, Eq. (7.28) says the variance of a sum of two independent random variables is the sum of their individual variances.

We are able to handle more than two random variables in the same way. As is known from before, if $X_1, X_2, \ldots, X_r$ are $r \geq 2$ random variables on Ω, then $E(\sum_{i=1}^{r} X_i) = \sum_{i=1}^{r} E(X_i)$. Let us denote $E(X_i)$ by μ_i, $i = 1, 2, \ldots, r$, and consider what might be said about $V(\sum_{i=1}^{r} X_i)$.

Proposition 7.8 $V(\sum_{i=1}^{r} X_i) = \sum_{i=1}^{r} V(X_i) + \sum_{\substack{i=1 \\ i \neq j}}^{r} \sum_{j=1}^{r} \text{Cov}(X_i, X_j)$. (The last sum extends over all pairs i and j for which $i \neq j$.)

PROOF: We have

$$
\begin{aligned}
V(\sum_{i=1}^{r} X_i) &= E((\sum_{i=1}^{r} X_i - \sum_{i=1}^{r} \mu_i)^2) \\
&= E((\sum_{i=1}^{r} (X_i - \mu_i))^2) \\
&= E(\sum_{i=1}^{r} (X_i - \mu_i)^2 + \sum_{\substack{i=1 \\ i \neq j}}^{r} \sum_{j=1}^{r} (X_i - \mu_i)(X_j - \mu_j)) \qquad (7.29) \\
&= \sum_{i=1}^{r} E((X_i - \mu_i)^2) + \sum_{\substack{i=1 \\ i \neq j}}^{r} \sum_{j=1}^{r} E((X_i - \mu_i)(X_j - \mu_j) \\
&= \sum_{i=1}^{r} V(X_i) + \sum_{\substack{i=1 \\ i \neq j}}^{r} \sum_{j=1}^{r} \text{Cov}\,(X_i, X_j)
\end{aligned}
$$

Equation (7.29) simplifies greatly if the random variables $X_1, \ldots, X_r$ are such that Cov $(X_i, X_j) = 0$ for all pairs i, j such that $i \neq j$. This is the case, for example, if the random variables $X_1, \ldots, X_r$ are pairwise independent (see Exercise 9 of Section 5.5). In that case, we get the following corollary.

Corollary 7.2 If $X_1, X_2, \ldots, X_r$ are pairwise independent random variables,

$$V(\sum_{i=1}^{r} X_i) = \sum_{i=1}^{r} V(X_i) \qquad (7.30)$$

Of course, Eq. (7.30) remains in effect if $X_1, X_2, \ldots, X_r$ are also mutually independent.

Example 7.7

In Example 7.3, we found the expectation of the honor count in a bridge hand. Suppose we now try to find the variance of the honor count knowing full well that a direct approach through Eq. (7.23) is not feasible. The honor count X has been expressed as a sum of 16 random variables, and Proposition 7.8 appears to be relevant. Specifically,

$$X = 4 \sum_{j=1}^{4} I_j + 3 \sum_{j=5}^{8} I_j + 2 \sum_{j=9}^{12} I_j + \sum_{j=13}^{16} I_j$$

In this sum, each indicator random variable relates to the specification of a single card, and the variance of each may be found to be $\frac{1}{4} \cdot \frac{3}{4} = \frac{3}{16}$. Applying (2) of Proposition 7.6, we see that the sum of the variances of the component parts of X contributes an amount $4(16 \cdot \frac{3}{16}) + 4(9 \cdot \frac{3}{16}) + 4(4 \cdot \frac{3}{16}) + 4(1 \cdot \frac{3}{16}) = \frac{45}{2}$ to $V(X)$.

We also require certain covariances. It will suffice to find the covariance between any choice of I_i and I_j, each of these relating to the specification of a single card in the hand. This may be found to be $-\frac{1}{272}$. If we then verify that $\text{Cov}\,(cI_i, dI_j) = c \cdot d \cdot \text{Cov}\,(I_i, I_j)$ (see Exercise 5), we are reduced to counting up the number of contributions of $-\frac{1}{272}$ that we have. Without going through the details, we claim that there are 1480 of them. Then

$$V(X) = \frac{45}{2} - \frac{1480}{272} = \frac{4550}{272} \sim 17$$

and σ_X is slightly larger than 4.

EXERCISES 7.4

1. Find the variance of the total score in two independent throws of a balanced die.

2. Consider a random sample of size 13 drawn without replacement from a standard deck of 52 cards. Find the variance of the following random variables.
(**a**) The number of aces drawn.
(**b**) The number of black aces drawn.
(**c**) The number of black cards drawn.
(**d**) The number of aces and face cards drawn.
(Use indicator random variables and the results of Example 7.7.)

3. Find the variance of an indicator random variable. In particular, show that $V(c) = 0$. [Compare with Eq. (7.10).]

4. Find the variance of the number of heads in four independent tosses of a coin. With no further computation, find also the variance of the number of tails and the covariance of the number of heads and the number of tails. (HINT: The variance of the sum of the number of heads and the number of tails is zero.)

5. Show that $\text{Cov}\,(cX, dY) = cd\,\text{Cov}\,(X, Y)$ and that $\text{Cov}\,(X + c, Y + d) = \text{Cov}\,(X, Y)$.

6. Suppose that X and Y are two random variables and that each assumes only two values. Show then that $\text{Cov}\,(X, Y) = 0$ implies X and Y are independent.

7. Show that if $V(X) = 0$ then $P(X = E(X)) = 1$. (See Exercise 15 of Section 7.2.)

8. Show that in Example 7.6, $V(X_n) = 1$ for all n (let I_j be the indicator random variable of the event that there is a match on the jth play, $j = 1, 2, \ldots, n$, and note that $X_n = \sum_{j=1}^{n} I_j$).

9. Show that $E((X - c)^2)$ is a minimum in c when $c = \mu = E(X)$. [HINT: Write $E((X - c)^2) = E(((X - \mu) + (\mu - c))^2)$ and expand the latter.]

10. A random variable is said to be *standardized* if its expectation is zero and its standard deviation (and hence variance) is one. Show that if X is a random variable with expectation μ and standard deviation $\sigma > 0$, then the random variable $(X - \mu)/\sigma$ is standardized.

11. Let $X_1, X_2, \ldots, X_n$ be random variables each with expectation μ and standard deviation σ. Suppose further that Cov $(X_i, X_j) = 0$ for $i \neq j$. If $\bar{X} = (X_1 + X_2 + \cdots + X_n)/n$, find $E(\bar{X})$ and $V(\bar{X})$ in terms of μ and σ^2.

12. With the same assumptions as in Exercise 11, show that

$$E\left(\sum_{i=1}^{n} (X_i - \bar{X})^2\right) = (n - 1)\sigma^2$$

[HINT: Write

$$E\left(\sum_{i=1}^{n} (X_i - \bar{X})^2\right) = E\left(\sum_{i=1}^{n} ((X_i - \mu) - (\bar{X} - \mu))^2\right)$$

and expand the latter.]

13. Let X and Y be two random variables on Ω, and let P be a probability measure on Ω. By noting that for any real number c, $E((cX + Y)^2) \geq 0$, deduce that $(E(XY))^2 \leq E(X^2)E(Y^2)$. This result is known as the *Cauchy-Schwarz inequality*. [HINT: $E((cX + Y)^2) = c^2E(X^2) + 2cE(XY) + E(Y^2)$, and the minumum value of this function of c must be nonnegative.]

14. Let X and Y be two random variables on Ω, and let P be a probability measure on Ω. Let μ_X, μ_Y be the expectations and σ_X, σ_Y be the positive standard deviations of X and Y, respectively. The *correlation coefficient* of X and Y, denoted by $\rho(X, Y)$, is defined by $\rho(X, Y) = \text{Cov}(X, Y)/\sigma_X\sigma_Y$. Show that $(\rho(X, Y))^2 \leq 1$. (HINT: Apply the inequality of Exercise 13 to the random variables $X - \mu_X$ and $Y - \mu_Y$.)

15. Show that if X is a random variable and $Y = cX + d$, $c \neq 0$, then Cov $(X, Y) = c\sigma_X^2$ and $\sigma_Y = |c|\sigma_X$. In this case, it may be seen that $\rho(X, Y) = \pm 1$ depending on whether $c > 0$ or $c < 0$.

7.5 Application to the binomial and hypergeometric distributions

In this section, we will find the expectation and variance for the binomial and hypergeometric distributions. They are given in Eqs. (7.34), (7.38), (7.48), and (7.53).

Let us begin with the binomial distribution. Thus, let S_n be a random variable whose distribution is binomial with parameters n and p. This means that S_n can take the values 0, 1, 2, . . . , n, and the probability that S_n equals r is given by

$$f_{S_n}(r) = \binom{n}{r} p^r q^{n-r}, \qquad 0 \leq r \leq n, \quad q = 1 - p \tag{7.31}$$

It follows from Eq. (7.8) that

$$E(S_n) = \sum_{r=0}^{n} r f_{S_n}(r) = \sum_{r=0}^{n} r \binom{n}{r} p^r q^{n-r} \tag{7.32}$$

This sum may be performed as follows. First, note that the term corresponding to $r = 0$ is zero, so the sum may be extended from $r = 1$ to $r = n$. Next, note that $r\binom{n}{r} = n\binom{n-1}{r-1}$. Then

$$E(S_n) = \sum_{r=1}^{n} n \binom{n-1}{r-1} p^r q^{n-r} = np \sum_{r=1}^{n} \binom{n-1}{r-1} p^{r-1} q^{n-r} \tag{7.33}$$

The sum on the right side of Eq. (7.33) is the binomial expansion of $(q + p)^{n-1}$. But $q + p = 1$, and therefore this sum must equal one. So,

$$E(S_n) = np \tag{7.34}$$

We may similarly compute $E(S_n^2)$ as follows.

$$\begin{aligned} E(S_n^2) &= \sum_{r=0}^{n} r^2 \binom{n}{r} p^r q^{n-r} = \sum_{r=0}^{n} [r(r-1) + r] \binom{n}{r} p^r q^{n-r} \\ &= \sum_{r=0}^{n} r(r-1) \binom{n}{r} p^r q^{n-r} + \sum_{r=0}^{n} r \binom{n}{r} p^r q^{n-r} \end{aligned} \tag{7.35}$$

The last sum in Eq. (7.35) is precisely $E(S_n) = np$. The other sum in this expression for $E(S_n^2)$ may be extended from $r = 2$ to $r = n$, because the terms corresponding to $r = 0$ and $r = 1$ are both zero. Moreover, $r(r-1)\binom{n}{r} = n(n-1)\binom{n-2}{r-2}$. Thus,

$$\begin{aligned} E(S_n^2) &= \sum_{r=2}^{n} n(n-1) \binom{n-2}{r-2} p^r q^{n-r} + np \\ &= n(n-1)p^2 \sum_{r=2}^{n} \binom{n-2}{r-2} p^{r-2} q^{n-r} + np \end{aligned} \tag{7.36}$$

Now the last summation in Eq. (7.36) is the binomial expansion of $(q + p)^{n-2}$, and, since $q + p = 1$, this sum must equal one, and we have

$$\begin{aligned} E(S_n^2) &= n(n-1)p^2 + np = n^2p^2 + np - np^2 \\ &= n^2p^2 + np(1-p) = n^2p^2 + npq \end{aligned} \tag{7.37}$$

Finally, $V(S_n) = E(S_n^2) - \mu^2$, and so

$$V(S_n) = n^2p^2 + npq - n^2p^2 = npq \tag{7.38}$$

We shall now give an alternative method for computing $E(S_n)$ and $V(S_n)$. This method is of considerable interest because it is applicable in other situations as well. The method proceeds by expressing S_n as a sum of rather simple random variables—indeed, as a sum of indicator random variables.

Imagine n independent trials of an experiment $\mathcal{E}$ with two outcomes S and F. As usual, we let $\Omega = \{S, F\}$. Let P be defined on Ω by $P(\{S\}) = p$. Then if S_n is the total number of successes in the n trials on $\mathcal{E}$, we have seen earlier that S_n is a random variable on Ω^n whose distribution is binomial with parameters n and p.

Now, for each integer i between 1 and n, we define a random variable X_i on Ω^n as follows:

$$\begin{aligned} X_i(\omega) &= 1 \text{ if } \omega \text{ has } S \text{ in the } i\text{th position.} \\ &= 0 \text{ if } \omega \text{ has } F \text{ in the } i\text{th position.} \end{aligned} \tag{7.39}$$

The random variable X_i is easily described verbally; it is just the number of successes in the ith trial of $\mathcal{E}$. (Note that this explains why X_i can only take the values zero or one.) Clearly, X_i is an indicator random variable, and the event $X_i = 1$ is exactly the event "Success in the ith trial."

The random variable S_n can be expressed in terms of $X_1, \ldots, X_n$. Indeed, since S_n is the total number of successes in the n trials of $\mathcal{E}$ and X_i is the number of successes in the ith trial, it follows that

$$S_n = X_1 + X_2 + \cdots + X_n \tag{7.40}$$

The random variable X_i depends only on the outcome of the ith trial of $\mathcal{E}$. Since the outcomes of the various trials are independent, it follows that $X_1, \ldots, X_n$ are mutually independent random variables. We have already so remarked in Chapter Six. Using Eq. (7.40), we now get

$$\begin{aligned} E(S_n) &= E(X_1) + \cdots + E(X_n) \\ V(S_n) &= V(X_1) + \cdots + V(X_n) \end{aligned} \tag{7.41}$$

Note that to get the first equation, the independence of $X_1, \ldots, X_n$ is not needed. Only Eq. (7.40) is needed. To get the second equation, we have used the mutual independence of $X_1, \ldots, X_n$, although, by virtue of Corollary 7.2, pairwise independence of $X_1, \ldots, X_r$ would have been sufficient.

In order to compute $E(S_n)$ and $V(S_n)$ now, we need only compute $E(X_i)$ and $V(X_i)$ for each i. We can do this easily by computing the probability distribution of X_i. $f_{X_i}(1)$ is the probability of the event $X_i = 1$. This event, as we have seen, is just the event "Success in the ith trial" and therefore has probability p. Thus, we get

$$f_{X_i} = p, \qquad f_{X_i}(0) = q \tag{7.42}$$

Immediately, we see that $E(X_i) = 1 \cdot f_{X_i}(1) + 0 \cdot f_{X_i}(0) = p$, and $E(X_i^2) = 1^2 \cdot f_{X_i}(1) + 0^2 \cdot f_{X_i}(0) = p$. Therefore, $V(X_i) = p - p^2 = pq$. Thus, each term on the right in the first equation in (7.41) equals p, and each term on the right in the second equation equals pq. Hence,

$$\begin{aligned} E(S_n) &= np \\ V(S_n) &= npq \end{aligned} \tag{7.43}$$

as before.

You must have noticed that each of the random variables $X_1, \ldots, X_n$ has the *same* distribution. The success of the above method is a result of the fact that the expectations and variances of the X_i can be computed without any trouble.

We shall now carry out the computation of the expectation and the variance of a random variable X whose distribution is hypergeometric with parameters N, k, n. We discussed this distribution in Section 6.5, so let us recall the setting. A random sample of size n is drawn without replacement from an urn containing N balls numbered $1, 2, \ldots, N$. Letting $\Omega = \{1, 2, \ldots, N\}$ we saw that $\Omega^{(n)}$ is an appropriate sample space and the equally likely measure P is the appropriate measure on $\Omega^{(n)}$. For a fixed integer k, we then considered the random variable X described as the number of balls among the n balls drawn, whose numbers are between 1 and k. It was found that X has the distribution described in Section 6.5, which we called the hypergeometric distribution with parameters N, k, n.

$E(X)$ and $V(X)$ can be computed by directly using the explicit formulas for f_X given in Section 6.5. However, that procedure is rather cumbersome, and we shall adopt an alternative method, which proceeds by expressing X as a sum of indicator random variables. For this purpose, for each integer i between 1 and n, define X_i on $\Omega^{(n)}$ by

$$\begin{aligned} X_i(\omega) &= 1 \text{ if } \omega \text{ has one of the integers } 1, 2, \ldots, k \text{ in the} \\ &\qquad i\text{th position.} \\ &= 0 \text{ otherwise.} \end{aligned} \tag{7.44}$$

Thus, $X_i = 1$ if the ith draw yields a ball numbered between 1 and k, and $X_i = 0$ otherwise. It follows that exactly as many X_i will take the value 1 as there are balls drawn whose numbers are between 1 and k. Thus,

$$X = X_1 + X_2 + \cdots + X_n \tag{7.45}$$

and we have expressed X as a sum of indicator random variables.

While Eq. (7.45) is analogous to (7.40), we should note that there is an important difference between them. In Eq. (7.40), the random variables $X_1, \ldots, X_n$ were mutually independent. In Eq. (7.45), they are not mutually or even pairwise independent, since drawings without replacement constitute dependent trials. When we want to express $E(X)$ in terms of $E(X_1), \ldots, E(X_n)$, this fact causes no complication, for we have

$$E(X) = E(X_1) + E(X_2) + \cdots + E(X_n) \tag{7.46}$$

whether or not $X_1, \ldots, X_n$ are independent. However, for computing the variance $V(X)$, we now have to take into consideration the covariances of X_i, X_j, and we get, by Eq. (7.45) and Proposition 7.9,

$$V(X) = \sum_{i=1}^{n} V(X_i) + \sum_{\substack{i=1 \\ i \neq j}}^{n} \sum_{j=1}^{n} \operatorname{Cov}(X_i, X_j) \tag{7.47}$$

To finish our task, we must now compute $E(X_i)$, $V(X_i)$, and Cov (X_i, X_j). We begin by computing the probability distribution of X_i for a fixed i. X_i takes the value 1 if and only if the ith ball drawn bears a number between 1 and k. Thus, the event $\{\omega | X_i(\omega) = 1\}$ in $\Omega^{(n)}$ consists of precisely those n-tuples in $\Omega^{(n)}$ whose ith coordinate is between 1 and k. To find the probability of this event, we merely count the number of such n-tuples and divide by the total number of points in $\Omega^{(n)}$. (Recall that $\Omega^{(n)}$ carries the equally likely measure.) To count the number of such n-tuples, we observe that the ith coordinate of such an n-tuple may be chosen in k different ways. Having chosen the ith coordinate, the rest of the coordinates of such an n-tuple may be filled in $(N-1)_{n-1}$ different ways, since $N-1$ balls are now available from which to fill the $n-1$ remaining coordinates of such an n-tuple. Thus, the total number of n-tuples in $\Omega^{(n)}$ that have the ith coordinate between 1 and k is $k(N-1)_{n-1}$. Hence, the probability of the event $X_i = 1$ is

$$\frac{k(N-1)_{n-1}}{(N)_n} = \frac{k}{N}$$

Thus,

$$f_{X_i}(1) = \frac{k}{N}, \qquad f_{X_i}(0) = 1 - \frac{k}{N} \tag{7.48}$$

It follows immediately that

$$E(X_i) = 1 \cdot f_{X_i}(1) + 0 \cdot f_{X_i}(0) = \frac{k}{N} \tag{7.49}$$

and

$$V(X_i) = \frac{k}{N} - \frac{k^2}{N^2} = \frac{k}{N}\left(1 - \frac{k}{N}\right) \tag{7.50}$$

Finally, we have to compute Cov (X_i, X_j), $i \neq j$. Since, by Eq. (7.26), Cov $(X_i, X_j) = E(X_iX_j) - E(X_i)E(X_j)$ and we already know $E(X_i)$ for each i by Eq. (7.49), we are led to compute $E(X_iX_j)$. Now, since X_i takes only the values 0 and 1 and so also X_j, it follows that $X_i \cdot X_j$ also takes only the values 0 and 1. Indeed, the event $X_iX_j = 1$ is exactly the event "$X_i = 1$ and $X_j = 1$." Therefore, the probability of $X_iX_j = 1$ is precisely the probability of the event "$X_i = 1$ and $X_j = 1$." X_i and X_j both take the value 1 if and only if *both* the ith and the jth draws result in balls bearing numbers between 1 and k. Thus, the event $\{\omega | X_i(\omega) = 1$ and $X_j(\omega) = 1\}$ in $\Omega^{(n)}$ consists precisely of those n-tuples in $\Omega^{(n)}$ having integers between 1 and k in *both* the ith and the jth position. To count the number of such n-tuples, observe that the ith and jth coordinates of such an n-tuple may be made up in $k(k - 1)$ ways and the remainder in $(N - 2)_{n-2}$ ways. Thus, the probability of this event is

$$\frac{k(k - 1)(N - 2)_{n-2}}{(N)_n} = \left(\frac{k}{N}\right)\left[\frac{(k - 1)}{(N - 1)}\right]$$

Thus,

$$f_{X_iX_j}(1) = \frac{k}{N} \cdot \frac{(k - 1)}{(N - 1)}, \qquad f_{X_iX_j}(0) = 1 - \frac{k}{N} \cdot \frac{(k - 1)}{(N - 1)} \tag{7.51}$$

and we find

$$E(X_iX_j) = \frac{k}{N} \cdot \frac{k - 1}{N - 1} \tag{7.52}$$

and

$$\begin{aligned} \text{Cov}\,(X_i, X_j) &= E(X_iX_j) - E(X_i)E(X_j) \\ &= \frac{k(k - 1)}{N(N - 1)} - \frac{k}{N} \cdot \frac{k}{N} \\ &= -\frac{k(N - k)}{N^2(N - 1)} \end{aligned} \tag{7.53}$$

We now return to Eq. (7.46). Using (7.49), we get

$$E(X) = \frac{nk}{N} \tag{7.54}$$

Finally, we see that in Eq. (7.47) each term of the type $V(X_i)$ is equal to $k(N - k)/N^2$ because of Eq. (7.50), and each term of the type

Cov (X_i, X_j) equals $-k(N-k)/N^2(N-1)$ by virtue of Eq. (7.53). There are n terms of the former type and $n(n-1)$ terms of the latter type. Hence,

$$V(X) = \frac{n \cdot k(N-k)}{N^2} - n(n-1) \cdot \frac{k(N-k)}{N^2(N-1)}$$

$$= \frac{nk(N-k)}{N^2} \cdot \frac{N-n}{N-1} \qquad (7.55)$$

terminating our computation.

It is of some interest to compare the expectation and variance of the hypergeometric distribution with the corresponding numbers for the binomial distribution. Here is why. Let us set $p = k/N$ in the case of the hypergeometric distribution. As we have seen in Eq. (7.48), p is equal to the probability of getting on a given draw a ball numbered between 1 and k. Now, suppose that instead of drawing the balls without replacement, we *replace* each ball, after noting its number, and before drawing the next ball. If we decide to say that getting a number between 1 and k is a success, then we will have Bernoulli trials with probability p of success in a given trial. Thus, when we draw the balls without replacement, the number of "successes" is a hypergeometric random variable, while if we draw the balls with replacement, the total number of "successes" is a binomial random variable. A comparison of these two distributions will therefore bring to light differences between the two methods of drawing. Reference to the expectations of these two distributions shows that the expectations of both are the same (i.e., equal to np where $p = k/N$), while the variance of the hypergeometric distribution is $npq \cdot [(N-n)/(N-1)]$, where $p = k/N$. This is a little less than npq, which is the variance of the binomial distribution. Thus, when we draw without replacement, the expected number of successes is the same as when we draw with replacement; however, the variance of the number of successes is smaller in the former case than in the latter.

EXERCISES 7.5

1. If a machine produces defective items with a probability of $\frac{1}{20}$, what is the expected number of defective items in a random sample of size 500 taken from its output? What is the variance of the number of defective items?

2. Let Y be the *proportion* of successes in n Bernoulli trials with a common probability p of success in a given trial. Find $E(Y)$ and $V(Y)$.

3. A certain drug is known to have a cure rate of one in four. If 200 patients are given the drug, what is the expectation of the *proportion* of cures? What is its variance?

4. Let X_i, $i = 1, \ldots, n$ be the number of successes in the ith trial in n Bernoulli trials with parameter p.
(**a**) For any positive integer r, find $E(X_i^r)$.
(**b**) Let r, s be any integers. Are X_i^r and X_j^s independent when $i \neq j$? Why?
(**c**) Compute $E(X_i^r X_j^s)$ where r, s are positive integers and $i \neq j$.
(**d**) Let $S_n = X_1 + X_2 + \cdots + X_n$, so that X is the total number of successes in the n trials. Using (**a**)–(**c**), find $E(S_n^3)$.
(**e**) Write an expression for $E(S_n^3)$ by using the probability distribution of S_n. Can you sum this expression?

5. If n independent multinomial trials are performed, and if X_i is the total number of outcomes of the ith kind, $i = 1, 2, \ldots, N$, find $E(X_i)$ and $V(X_i)$ in terms of n and the parameters $p_1, p_2, \ldots, p_N$.

6. In Exercise 5, represent X_i as a sum of n independent random variables. What is the distribution of each of these?

7. By using the representation of X_i as the sum of n independent random variables and by doing the same thing for X_j, find Cov (X_i, X_j) in the setting of Exercise 5. Are X_i and X_j independent? Why?

8. Let $N = 3$ in Exercise 5. Thus, we have a trinomial distribution with parameters p_1, p_2, p_3. Let X_i be the total number of outcomes of the ith kind $(i = 1, 2, 3)$ when n such independent trinomial experiments are performed. Find $E(X_1)$, $V(X_1)$, and Cov (X_1, X_2) by direct computation using the distribution of X_1 and the joint distribution of X_1 and X_2.

9. Suppose 20 independent tosses of a balanced die are made. Find the covariance between the number of ones and the number of sixes that appear.

10. A random variable X that takes each of values $1, 2, \ldots, N$ with the same probability $1/N$ is said to have a *uniform distribution on* $1, 2, \ldots, N$. Let $X_1, X_2, \ldots, X_n$ be n mutually independent random variables each with a uniform distribution on $1, 2, \ldots, N$.
(**a**) Find $E(X_i)$ and $V(X_i)$.
(**b**) Let $X = X_1 + X_2 + \cdots + X_n$. Does X have a uniform distribution on $1, 2, \ldots, N$? On $1, 2, \ldots, nN$?
(**c**) Find $E(X)$ and $V(X)$.
(**d**) Let M_n be the largest of $X_1, X_2, \ldots, X_n$ and m_n be the smallest of $X_1, X_2, \ldots, X_n$. Find the probability distribution of M_n and of m_n. (HINT: $M_n \leq k$ if and only if *all* the random variables $X_1, X_2, \ldots, X_n$ take a value less than or equal to k. The event $M_n = k$ is just the event "$M_n \leq k$ and $M_n > k - 1$.")

11. Show that the waiting time distribution of the example of Section 6.5 (Example 6.6) is a uniform distribution on $1, 2, \ldots, N$ if N draws are made from N balls with $k = 1$. Hence, find the variance of the waiting time.

12. Let X be a random variable with a hypergeometric distribution with parameters N, k, n. Derive Eq. (7.54) by direct computation, without representing X as a sum of the random variables $X_1, \ldots, X_n$ as in the text.

13. A lake contains 1000 tagged fish and 5000 untagged fish. If a random sample of size 120 is drawn from the lake without replacement, what is the expectation of the number of tagged fish caught? What is the variance?

14. Let X be a random variable having a hypergeometric distribution with parameters N, k, and n. Show that if n is held fixed but N and k tend to infinity in such a way that $k/N = p$, then $E(X)$ tends to np and $V(X)$ tends to $np(1 - p)$. Explain why this phenomenon is not surprising.

7.6 The Chebychev inequality

This section is devoted to the derivation of an important inequality due to P. L. Chebychev (1821–94). This inequality is a somewhat more precise statement of the qualitative assertion that the variance measures the spread or dispersion of a probability distribution. Since $V(X) = E((X - \mu)^2)$ for a random variable X, it is evident that if $V(X)$ is small, then $(X - \mu)^2$ (or $|X - \mu|$) cannot be too large at too many points of Ω. Thus, the size of $V(X)$ limits the probability of large deviations of X from its expectation. Specifically, we prove the following proposition.

Proposition 7.9 *The Chebychev Inequality* Let P be a probability measure on a sample space Ω, and let X be a random variable on Ω having expectation μ and variance V. Then, for any positive number α,

$$P(|X - \mu| \geq \alpha) \leq \frac{V}{\alpha^2} \tag{7.56}$$

PROOF: Let A be the event of Ω defined by $A = \{\omega \mid |X(\omega) - \mu| \geq \alpha\}$. Then the left side of Eq. (7.56) is precisely $P(A)$. Now, denoting $V(X)$ by V,

$$\begin{aligned} V = V(X) &= E((X - \mu)^2) \\ &= \sum_{\omega \in \Omega} (X(\omega) - \mu)^2 P(\{\omega\}) \end{aligned} \tag{7.57}$$

In this last sum, let us collect together the terms for which $\omega \in A$ and those for which $\omega \in A'$. Then

$$V = \sum_{\omega \in A} (X(\omega) - \mu)^2 P(\{\omega\}) + \sum_{\omega \in A'} (X(\omega) - \mu)^2 P(\{\omega\}) \tag{7.58}$$

The second sum here is nonnegative so that

$$V \geq \sum_{\omega \in A} (X(\omega) - \mu)^2 P(\{\omega\}) \tag{7.59}$$

Now for each $\omega \in A$, $|X(\omega) - \mu| \geq \alpha$ by the definition of A. Hence, for each $\omega \in A$, $(X(\omega) - \mu)^2 \geq \alpha^2$. So,

$$\begin{aligned} V &\geq \sum_{\omega \in A} (X(\omega) - \mu)^2 P(\{\omega\}) \geq \sum_{\omega \in A} \alpha^2 P(\{\omega\}) \\ &= \alpha^2 \sum_{\omega \in A} P(\{\omega\}) = \alpha^2 P(A) \end{aligned} \tag{7.60}$$

In other words,

$$P(A) \leq \frac{V}{\alpha^2} \tag{7.61}$$

which is precisely the assertion of the Chebychev inequality.

The main virtue of Proposition 7.9 is that it requires no assumptions about the random variable X. However, a price is paid for this generality, namely, the estimate that Eq. (7.56) gives for the probability that $|X - \mu| \geq \alpha$ may not be very good. That is, the actual value of $P(|X - \mu| \geq \alpha)$ may be considerably less than $V(X)/\alpha^2$ and then the value $V(X)/\alpha^2$ is a crude overestimation of $P(|X - \mu| \geq \alpha)$. The usefulness of the Chebychev inequality is mainly as a theoretical tool rather than as a practical device for estimating probabilities like $P(|X - \mu| \geq \alpha)$. Its principal use for us will appear in Chapter Eight.

In conclusion, we give an alternative form of Eq. (7.56). Since (7.56) is true for any $\alpha > 0$, it is true for $\alpha = \beta\sigma_X$ where $\beta > 0$ and σ_X is the (positive) standard deviation of $X(\sigma_X^2 = V(X))$. With this choice, Eq. (7.56) becomes

$$P(|X - \mu| \geq \beta\sigma_X) \leq \frac{1}{\beta^2} \tag{7.62}$$

In words, the probability that X deviates from its expectation by at least β standard deviations is no more than $1/\beta^2$. For instance, the probability that a random variable deviates from its expectation by at least three standard deviations is no more than $\frac{1}{9}$.

EXERCISES 7.6

1. If X is a binomial random variable with parameters 4 and $\frac{1}{2}$, compare $P(|X - \mu| \geq \alpha)$ with $V(X)/\alpha^2$ for $\alpha = \frac{1}{2}, 1$ and $\frac{3}{2}$. Repeat when X is instead a binomial random variable with parameters 4 and $\frac{1}{8}$.

2. Suppose X takes the values $0, 1, 2, \ldots, n$, and that $E(X) = 1$, $V(X) = 1$. Show that $P(X \geq 3) \leq \frac{1}{4}$. In the same way show that for $k = 4, 5, \ldots, n$, $P(X \geq k) \leq 1/(k-1)^2$.

3. Show that $P(|X - \mu| \geq \alpha) = V(X)/\alpha^2$ when X has a probability distribution given by $f_X(\alpha) = \frac{1}{2} = f_X(-\alpha)$. Can you see why this should be so in the proof of the Chebychev inequality?

4. Let X have a uniform distribution on 1, 2, 3, 4, 5. Graph the following two functions of $\alpha > 0$: $P(|X - \mu| \geq \alpha)$ and $V(X)/\alpha^2$. Comment on the meaning of this relative to inequalities on $P(|X - \mu| \geq \alpha)$.

5. Show that at least one of the numbers $P(X - \mu \geq \alpha)$ and $P(X - \mu \leq -\alpha)$ is no larger than $V(X)/2\alpha^2$.

6. Apply the Chebychev inequality to the random variable $\bar{X}$ in the setting of Exercise 11 of Section 7.4.

7. Show that if X is a random variable with expectation μ and if α is a positive number, then $P(|X - \mu| \geq \alpha) \leq E((X - \mu)^4)/\alpha^4$. Graph the following two functions of $\alpha > 0$ when X has a uniform distribution on 1, 2, 3, 4, 5: $V(X)/\alpha^2$ and $E((X - \mu)^4)/\alpha^4$. Comment on the meaning of this relative to inequalities on $P(|X - \mu| \geq \alpha)$.

CHAPTER EIGHT

Sums of Independent Random Variables

8.1 Introduction

In this chapter, we will study two fundamental theorems ot probability theory. The first, the so-called weak law of large numbers, we prove in Section 8.3. The second theorem, which is known as the central limit theorem, we state (but do not prove) in Section 8.5. A proof of the latter result is outside the scope of this book.

The results named above are not only basic to the theory of probability per se, but they have at the same time a multitude of important applications. For this reason, a major portion of the present chapter is given over to a discussion of applications.

8.2 Independent and identically distributed random variables

In applications of probability theory, one is often led to consider a collection of several mutually independent random variables, each of which has the same probability distribution. Such a collection of random variables is said simply to be independent and identically distributed. Later in this section we will give some examples of important practical situations to which the study of independent and identically distributed random variables is directly relevant. Before turning to these examples, however, we will sketch briefly the manner in which such random variables generally appear in a more theoretical framework. This will, in its own way, point toward the applications we have in mind.

Let us begin by imagining some experiment $\mathcal{E}$ for which the probability measure P on the sample space Ω is a suitable mathematical model. Suppose that X is a random variable on Ω. Then X may be thought of as some quantity or measurement whose value depends on the outcome of $\mathcal{E}$.

Next, suppose that n replications of $\mathcal{E}$ are performed independently and under identical conditions. We have seen that Ω^n is an appropriate sample space for the composite experiment $\mathcal{E}^n$ and that the product probability measure P^n on Ω^n embodies the independence assumptions concerning the separate trials of $\mathcal{E}$.

Now, each replication of $\mathcal{E}$ will yield a value of X. Let us take X_i to be the value of X yielded by the ith replica of $\mathcal{E}$. In other words, X_i is the value of the measurement X in the ith trial when n independent trials of $\mathcal{E}$ are performed. X_i may be thought of as a random variable on Ω^n for each of $i = 1, 2, \ldots, n$. However, it is a special kind of random variable on Ω^n in that its value depends only on the outcome of the ith trial. More precisely, if $\omega = (\omega_1, \omega_2, \ldots, \omega_n)$ is a sample point of Ω^n, we have

$$X_i(\omega) = X(\omega_i), \qquad i = 1, 2, \ldots, n \tag{8.1}$$

Since the ith trial of $\mathcal{E}$ may result in any outcome in Ω, it follows that X_i can assume any of the values that X can assume. Furthermore, if x is any such value, the definition of P^n insures that

$$P^n(X_i = x) = P(X = x), \qquad i = 1, 2, \ldots, n \tag{8.2}$$

The full implication of Eq. (8.2) is that each of the random variables $X_1, X_2, \ldots, X_n$ has the same probability distribution as does X. Moreover, since X_i depends only on the outcome of the ith trial of $\mathcal{E}$ and since the trials of $\mathcal{E}$ are independent, $X_1, X_2, \ldots, X_n$ are in fact mutually independent (see Section 6.2 and Exercise 12 following it).

We see that we have above random variables $X_1, X_2, \ldots, X_n$ on Ω^n, which are independent and identically distributed. Note that we have not specified what their common probability distribution is. Indeed, it can be any probability distribution whatsoever, fixed once and for all by the choice of X.

Now we illustrate some situations for which the above abstraction is appropriate. You will, of course, recognize some old friends.

Example 8.1

Consider n independent trials of an experiment $\mathcal{E}$ having only two outcomes, say S and F. We remarked in Section 6.3 on the fact that such Bernoulli trials pertain to a variety of practical situations of frequent occurrence

We take $\Omega = \{S, F\}$ and suppose P is the probability measure determined on Ω by $P(\{S\}) = p$. Let X be the random variable defined on Ω by $X(S) = 1$ and $X(F) = 0$. Thus, X is the number of successes obtained in a single performance of $\mathcal{E}$, and X takes the values 1 and 0 with probabilities p and q, respectively.

For the experiment $\mathcal{E}^n$ consisting of n independent trials of $\mathcal{E}$, we let X_i be the number of successes obtained in the ith trial, $i = 1, 2, \ldots, n$. As random variables on the sample space Ω^n endowed with the product probability measure P^n, we see from Section 6.3 that $X_1, X_2, \ldots, X_n$ are mutually independent and that each of these takes the values 1 and 0 with probabilities p and q, respectively. Note that X_i and X satisfy Eq. (8.1) so that this is a special case of the setup described above.

Example 8.2

A physicist makes measurements of a quantity, say the velocity of light. It is a fact of experience that such measurements are subject to random errors, i.e., the errors are not systematic. Thus, when the physicist performs his experiment (let us call it $\mathcal{E}$), we may assume that his measurement yields one of finitely many values $c_1, c_2, \ldots, c_k$ for the velocity of light. (Of course, we would expect these values to cluster around the actual value, say c, of the velocity of light.)

We formalize this situation by supposing the measurement of the velocity of light obtained from one experiment is a random variable X with a probability distribution given by $f_X(c_j) = p_j$, $j = 1, 2, \ldots, k$. Thus, p_j is taken to be the probability that the error of measurement is $c_j - c$, $j = 1, 2, \ldots, k$. The probabilities $p_1, p_2, \ldots, p_k$ then characterize the way in which errors of measurement creep in. For our present purpose, we need not be specifically concerned with what these probabilities might be.

When quantities like the velocity of light are measured, it is a common practice to repeat the experiment, say n times. Further, these trials are performed independently under identical conditions as far as possible. If X_i is taken to be the velocity of light obtained from the ith trial of $\mathcal{E}$ under these idealized conditions, $i = 1, 2, \ldots, n$, then $X_1, X_2, \ldots, X_n$ are mutually independent random variables each taking the values $c_1, c_2, \ldots, c_k$ with the probabilities $p_1, p_2, \ldots, p_k$, respectively.

Example 8.3

In studying some numerical characteristic of a population of N people—suppose it is age in years—a statistician randomly selects a person in order to determine his or her age. Let us call the experiment of selection

$\mathcal{E}$ and the age determined by the selection X. If, in the population, there are N_j people of age j, say $j = 0, 1, 2, \ldots, 125$, then a random selection gives X the probability distribution $f_X(j) = N_j/N, j = 0, 1, 2, \ldots, 125$. The probabilities $N_0/N, N_1/N, N_2/N, \ldots, N_{125}/N$ here characterize the age distribution within the population, and, although some information might be available about them, they are typically unknown.

To gain information about population characteristics like age, the experiment $\mathcal{E}$ may be performed a number of times independently and under the same conditions, say n times. This amounts to random sampling with replacement, i.e., the person selected at each stage is returned to the population and may possibly be drawn again. To the extent that random sampling is performed independently a total of n times, the age of the ith person selected, X_i, has the probability distribution f_X, and, further, the random variables $X_1, X_2, \ldots, X_n$ are mutually independent. Again we are led to the consideration of a family of independent and identically distributed random variables.

8.3 The weak law of large numbers

In our remarks in Chapter Three, we alluded to the following basic empirical fact: if n trials of an experiment are performed and the event A is observed to have occurred T_n times in the n trials, then, as n becomes large, the values of the relative frequencies T_n/n tend to cluster together. It is natural to ask whether the mathematical theory of probability contains an assertion that is the theoretical counterpart of this empirical fact. The weak law of large numbers, which will be our concern in this section, is the name given to one such assertion.

Imagine an experiment $\mathcal{E}$, a sample space Ω for the outcomes of $\mathcal{E}$, and an event A. Suppose also that we have agreed on a probability measure P on Ω. Let us write p for $P(A)$ and $q = 1 - p = P(A')$. Now suppose we perform n independent replications of $\mathcal{E}$ and observe how many times the event A occurs. Since our only interest in each trial is in whether or not A occurs, we idealize the situation by saying that we have here Bernoulli trials, where S (success) stands for the occurrence of A and F (failure) stands for the nonoccurrence of A (or the occurrence of A'). Clearly, the probability of a success in a given trial is p. Also, the number of times the event A occurs in the n trials is just S_n, the total number of successes in the n Bernoulli trials. Therefore, the relative frequency of occurrence of A in the n trials is precisely S_n/n. Thus, a theoretical counterpart for the empirical observation mentioned above takes the form of an assertion that as n becomes large the values of the random variable S_n/n must tend to cluster together in some sense around a fixed number. Further, intuition suggests that this number should be the probability of A, viz., p.

The earliest form of the weak law of large numbers is the assertion that in a sense the values of S_n/n do indeed tend to stabilize around p as $n \to \infty$. This assertion is known as *Bernoulli's theorem*.

Theorem 8.1 *Bernoulli's Theorem* Let S_n be the number of successes in n Bernoulli trials, and let p be the probability of success in any given trial. Let ϵ be any positive number. Then

$$P^n\left(\left|\frac{S_n}{n} - p\right| > \epsilon\right) \to 0 \quad \text{as} \quad n \to \infty \tag{8.3}$$

PROOF: Letting X_i be the number of successes in the ith trial, we see that

$$S_n = X_1 + X_2 + \cdots + X_n \tag{8.4}$$

Now since X_i takes the values 1 and 0 with probabilities p and q, we have $E(X_i) = p$. Thus, $E(S_n) = np$, and $E(S_n/n) = p$. Further, $V(S_n) = npq$, so that $V(S_n/n) = (1/n^2)V(S_n) = pq/n$. Applying the Chebychev inequality to the random variable S_n/n, we find

$$P^n\left(\left|\frac{S_n}{n} - E\left(\frac{S_n}{n}\right)\right| > \epsilon\right) \leq \frac{V(S_n/n)}{\epsilon^2} \tag{8.5}$$

or

$$P^n\left(\left|\frac{S_n}{n} - p\right| > \epsilon\right) \leq \frac{pq}{n\epsilon^2} \tag{8.6}$$

Clearly, as $n \to \infty$, the right side tends to zero, and the left side, being nonnegative, must tend to zero as well. This finishes the proof.

In words, Bernoulli's theorem states that as the number n of trials becomes large, the probability that the proportion of successes in n Bernoulli trials deviates from p by more than a *preassigned* number ϵ approaches zero. Thus, if n is large, we expect that with a high probability S_n/n will be near p. It is in this sense that the values of S_n/n tend to cluster together.

Before commenting further on this theorem and proceeding to some applications, let us state and prove a generalization of Bernoulli's theorem. In Bernoulli's theorem, S_n was the sum of the random variables $X_1, X_2, \ldots, X_n$, which were independent identically distributed random variables. However, the random variables X_i were of a very special kind, for they could only take the values 1 and 0. We may ask what happens if the random variables $X_1, X_2, \ldots, X_n$ are not of this simple type. More precisely, we consider the following question: if $X_1, X_2, \ldots, X_n$ are independent random variables with a common distribution (not necessarily of the simple type involved in Bernoulli's theorem), and if we set $S_n = X_1 + X_2 + \cdots + X_n$, then do the values of S_n/n tend to cluster around some fixed number as n becomes large?

In Bernoulli's theorem, we saw that the values of S_n/n cluster around p. Note that p is the expectation of each of the random variables X_i involved in the Bernoulli case. Following this analogy, we observe that, in general, if $X_1, X_2, \ldots, X_n$ are identically distributed, then they all have the same expectation, say μ; let us see if we can prove in the present case an assertion analogous to Bernoulli's theorem with μ playing the role of p. More precisely, if $X_1, X_2, \ldots, X_n$ are independent identically distributed random variables with a common expectation μ, and if $S_n = X_1 + X_2 + \cdots + X_n$, we ask whether in some sense we can assert that the values of S_n/n tend to stabilize around μ.

That such an assertion can be made is the content of the so-called *weak law of large numbers* (in deference, of course, to a different law, called the strong law of large numbers which we will not even state). To be definite, we adhere to the notation of Section 8.2. Thus, suppose X is a random variable on a sample space Ω on which P is a probability measure. We have shown, for each n, how to construct random variables $X_1, X_2, \ldots, X_n$ on Ω^n which are mutually independent with respect to the product probability measure P^n and which have each the same probability distribution as does X.

Theorem 8.2 *The Weak Law of Large Numbers* For each n, let $X_1, X_2, \ldots, X_n$ be independent and identically distributed random variables on Ω^n, and let $S_n = X_1 + X_2 + \cdots + X_n$. Then, for any $\epsilon > 0$,

$$P^n\left(\left|\frac{S_n}{n} - \mu\right| > \epsilon\right) \to 0 \quad \text{as} \quad n \to \infty \tag{8.7}$$

where μ is the expectation of each of the X_i.

PROOF: Since the X_i are identically distributed, they have the same expectation μ. Then $E(S_n) = n\mu$ so that $E(S_n/n) = \mu$. Next, the variances of the random variables $X_1, X_2, \ldots, X_n$ are all equal; let σ^2 be the common value. Then

$$\begin{aligned} V(S_n) &= V(X_1) + V(X_2) + \cdots + V(X_n) \\ &= \sigma^2 + \cdots + \sigma^2 \\ &= n\sigma^2 \end{aligned} \tag{8.8}$$

because $X_1, X_2, \ldots, X_n$ are mutually independent. Hence $V(S_n/n) = (1/n^2)V(S_n) = \sigma^2/n$. Now applying the Chebychev inequality to the random variable S_n/n, we get

$$P^n\left(\left|\frac{S_n}{n} - E\left(\frac{S_n}{n}\right)\right| > \epsilon\right) \leq \frac{V(S_n/n)}{\epsilon^2} \tag{8.9}$$

That is,

$$P^n\left(\left|\frac{S_n}{n} - \mu\right| > \epsilon\right) \leq \frac{\sigma^2}{n\epsilon^2} \tag{8.10}$$

It is clear that the right side tends to zero as $n \to \infty$, and therefore so does the left side. This completes the proof.

Thus, as n becomes large, the probability that the average S_n/n differs from the expected value μ by more than ϵ tends to zero. Therefore, we would expect that, as n is increased, the values of the random variable S_n/n would be likely to be near μ, its expectation.

A value of S_n/n is, of course, an average of the values of $X_1, X_2, \ldots, X_n$, and the weak law of large numbers states, roughly, that the larger the number of trials on which the average is based, the better the chance that the average value is near its expectation. One may say that the process of averaging somehow cancels random deviations in each of the X_i. When, in common parlance, we speak of the law of averages being operative in some situations, we are really talking about the tendency of averages to stabilize around a fixed value, as exemplified by the weak law of large numbers.

We shall conclude by pointing out how the weak law of large numbers gives a theoretical justification for some statistical practices based on intuition.

Example 8.4

We have seen repeatedly that Bernoulli trials are a useful model for a variety of practical situations. Let us take the example of a production line on which items are being produced in succession. In previous chapters, we started by assuming that the probability of an item being defective is p, where $0 < p < 1$, and that the successive items produced could be thought of as independent Bernoulli trials. Assuming we know what p is, our study of Bernoulli trials can answer such questions as, "What is the probability of getting r or more defectives when n items are produced?" or "What is the expected number of defectives among n items inspected?" Given such a production line in real life, we cannot say a priori what is an appropriate value for p. For, presumably, the value of p is a characteristic of the process of production and inspection. Therefore, we have to rely on past experience and estimate a value of p that may then be used to make predictions and/or judgments. Of course, if we are going to estimate p and use the estimated value for making predictions, our estimated value must at least be a good one (i.e., close to reality) or our predictions will be inaccurate even if our model is the right one.

To determine what p is, suppose we have observed the production process over a period of time, inspecting every item produced during this period. Suppose n items were inspected and S_n turned out to be defective. A statistician then customarily takes S_n/n as an estimate for the probability p of a given item being defective. Further, the larger n is, the more confident one tends to be that this estimate is reliable. For example, if we inspect a day's production and find that the proportion of defective items is .057, and if upon observing a month's production, we find that the proportion of defective items is .049, our intuition suggests that the second figure is more reliable, since it is based on more extended observation.

The weak law of large numbers is, in a way, a theoretical justification of this belief. (You may prefer to regard it as the theoretical fact that prompts this belief.) It says in this context that if n is large, the proportion S_n/n of defective items observed will very likely be near p. Thus, the above procedure of estimating p is supplied with a theoretical basis. At this point, you may want to ponder the comments on the frequency theory made in Chapter Three. In particular, does the weak law of large numbers *prove* the frequency theory in any sense?

Example 8.5

Let us consider again the physicist encountered in Example 8.2 who was making repeated measurements of the velocity of light. Let us suppose that he makes n such measurements, thus obtaining n values for the random variables $X_1, X_2, \ldots, X_n$. It is customary after making n such measurements to take their average $(X_1 + X_2 + \cdots + X_n)/n$ as an estimate of the true value c of the velocity of light. The reasoning here is that in the process of summing, the errors of opposite signs in $X_1, X_2, \ldots,$ X_n tend to cancel and thus the average $(X_1 + X_2 + \cdots + X_n)/n$ has a better chance of being error free.

The weak law of large numbers tells us that this average value, if n is large, is likely to be near the expectation of the random variable X. (Cf. Section 8.2. Recall that all the random variables $X_1, X_2, \ldots, X_n$ have the same distribution as that of X, and hence the same expectation.) We also commented in Section 8.2 that the probability distribution of X is characteristic of the random way in which the measurements are subject to experimental error. It is usual to assume that these errors are so distributed that the expectation of X is equal to the true value c of the velocity of light. This is, of course, equivalent to saying that the expectation of the errors is zero. Granting this, the weak law of large numbers says that $(X_1 + X_2 + \cdots + X_n)/n$ is likely to be near c, if n is large. This is, of course, a theoretical justification of the practice of taking $(X_1 + X_2 + \cdots + X_n)/n$ as an estimate of c.

In the next example we illustrate an application somewhat more detailed than the examples above.

Example 8.6

Let us consider again random sampling with replacement from the age distribution of a population as discussed in Example 8.3. The idealized sampling produces $X_1, X_2, \ldots, X_n$ which are independent and identically distributed. With the notation used before, each X_i has the probability distribution $f_X(j) = N_j/N, j = 0, 1, 2, \ldots, 125$. Therefore, each X_i has the expectation $\mu = \sum_{j=0}^{125} j(N_j/N)$. The number μ is also called the population average inasmuch as it is the number obtained by adding all ages and dividing by the number of people. (Why?)

If the purpose of the sampling is to obtain information about the population average μ, the weak law of large numbers tells us that the sample average $(X_1 + X_2 + \cdots + X_n)/n$ is likely to be near to μ when n is large. So, for large samples, we expect to find the sample average (which we can compute) close to the population average (which is unknown).

Is this method as useful as it appears? In particular, would a number n of samples significantly less than N suffice to bring the sample average close to the population average? If not, it might be as easy to sample N times without replacement and thereby determine μ exactly. We should even expect this to be the case when N is not very large. One clue to these considerations is given in Eq. (8.10), where we find a bound on the probability that the sample average deviates from the population average by more than some number ϵ. Let us see what meaning this might have.

Suppose we want to find an estimate for μ that will be accurate to within 1 year, i.e., $\epsilon = 1$. This cannot be *guaranteed* by sampling with replacement, for we might, for example, happen always to pick the oldest man. However, such occurrences are not likely, and, if a sample of size n is drawn, the probability that the sample and population averages differ by more than 1 year is no larger than σ^2/n. Unfortunately, σ^2 would not be known either, since it is the variance of any one of the X_i, viz., $\sum_{j=0}^{125} (j - \mu)^2(N_j/N)$. On the other hand, it may be possible to give a decent upper bound to σ^2, which is all that we would require for now.

For example, imagine the age distribution of all people in the United States. It is clear that σ^2 in this case is no more than $(125)^2$. We are even willing to make a quite uneducated guess that $\sigma^2 \leq 400$, i.e., the standard deviation in this age distribution is no more than 20 years. With such a bound on σ^2, the probability that the sample average deviates from the population average by more than 1 for a sample size n is no more than $400/n$. This is certainly not much use if $n \leq 400$. For $n = 40{,}000$, however, the probability is no more than $\frac{1}{100}$. Again, this is uninteresting if $N = 1000$, but, if N is approximately 200,000,000 as it would be for the population of the United States, 40,000 samples represents only .02% of the total population.

We assert that a random sample of 40,000 drawn with replacement from the population of the United States would give a sample average age deviating by more than 1 from the population average age no more than 1% of the time. This statement is certainly a useful assertion. However, our methods for obtaining it have been very crude. More refined methods may be brought to bear, which will improve on the claims above. In particular, we will discuss some improvements in Section 8.5.

EXERCISES 8.3

1. Suppose in a community there are 50,000 family units, and of these 35,000 own their dwellings and 15,000 do not. If a random sample of size 500 is drawn from these family units with replacement, what is the expectation and the variance of the sample proportion of family units owning their dwellings? Answer the same question in the case that the random sample is drawn without replacement.

2. Suppose in a community there are 100,000 adult males of whom 20,000 own no cars, 60,000 own one car, 15,000 own two cars, and 5000 own three cars. What is the population average number of cars? If a random sample of size 1000 is drawn from these adult males with replacement, what is the expectation and the variance of the sample average number of cars owned? Answer the last question in the case that the random sample is drawn without replacement.

3. The weak law of large numbers is proved under the assumption that the random variables being averaged are independent and identically distributed. Does this law hold if

(a) The random variables are pairwise independent and identically distributed?

(b) The random variables are pairwise independent and have each the same expectation and variance?

(c) The random variables have covariance 0 and are identically distributed?

(d) The random variables have covariance 0 and have each the same expectation and variance?

4. For each n, let $X_1, X_2, \ldots, X_n$ be independent and identically distributed with some fixed probability distribution having expectation μ and second moment μ_2. Show that for each $\epsilon > 0$,

$$P^n\left(\left|\frac{1}{n}\sum_{i=1}^{n} X_i^2 - \mu_2\right| > \epsilon\right) \to 0 \quad \text{as} \quad n \to \infty$$

5. Under the assumptions of Exercise 4, show that for each $\epsilon > 0$,

$$P^n\left(\left|\left(\frac{1}{n}\sum_{i=1}^{n} X_i\right)^2 - \mu^2\right| > \epsilon\right) \to 0 \quad \text{as} \quad n \to \infty$$

[HINT: Write $\left|\left(\frac{1}{n}\sum_{i=1}^{n} X_i\right)^2 - \mu^2\right| = \left|\frac{1}{n}\sum_{i=1}^{n} X_i - \mu\right|\left|\frac{1}{n}\sum_{i=1}^{n} X_i + \mu\right|$, and observe that $\left|\frac{1}{n}\sum_{i=1}^{n} X_i + \mu\right|$ is always smaller than some positive number M (Why?). Then the event that

$$\left|\left(\frac{1}{n}\sum_{i=1}^{n} X_i\right)^2 - \mu^2\right| = \left|\frac{1}{n}\sum_{i=1}^{n} X_i - \mu\right|\left|\frac{1}{n}\sum_{i=1}^{n} X_i + \mu\right| > \epsilon$$

is contained in the event that $\left|\frac{1}{n}\sum_{i=1}^{n} X_i - \mu\right| > \frac{\epsilon}{M}$ and the conclusion will follow from the weak law of large numbers.]

6. Under the assumptions of Exercise 4, show that for each $\epsilon > 0$,

$$P^n\left[\left|\frac{1}{n}\sum_{i=1}^{n} X_i^2 - \left(\frac{1}{n}\sum_{i=1}^{n} X_i\right)^2 - \sigma^2\right| > \epsilon\right] \to 0 \quad \text{as} \quad n \to \infty$$

where $\sigma^2 = \mu_2 - \mu^2$. [HINT: Notice that

$$\left|\frac{1}{n}\sum_{i=1}^{n} X_i^2 - \left(\frac{1}{n}\sum_{i=1}^{n} X_i\right)^2 - (\mu_2 - \mu^2)\right| \le \left|\frac{1}{n}\sum_{i=1}^{n} X_i^2 - \mu_2\right| + \left|\left(\frac{1}{n}\sum_{i=1}^{n} X_i\right)^2 - \mu^2\right|$$

and therefore that the event

$$\left|\frac{1}{n}\sum_{i=1}^{n} X_i^2 - \left(\frac{1}{n}\sum_{i=1}^{n} X_i\right)^2 - \sigma^2\right| > \epsilon$$

is contained in the event that the larger of

$$\left|\frac{1}{n}\sum_{i=1}^{n} X_i^2 - \mu_2\right| \quad \text{and} \quad \left|\left(\frac{1}{n}\sum_{i=1}^{n} X_i\right)^2 - \mu^2\right|$$

is more than $\epsilon/2$. The probability of this latter event is smaller than

$$P^n\left(\left|\frac{1}{n}\sum_{i=1}^{n} X_i^2 - \mu_2\right| > \frac{\epsilon}{2}\right) + P^n\left(\left|\left(\frac{1}{n}\sum_{i=1}^{n} X_i\right)^2 - \mu^2\right| > \frac{\epsilon}{2}\right)$$

(Why?).]

7. Show that

$$\frac{1}{n}\sum_{i=1}^{n}\left[X_i - \left(\frac{1}{n}\sum_{i=1}^{n} X_i\right)\right]^2 = \frac{1}{n}\sum_{i=1}^{n} X_i^2 - \left(\frac{1}{n}\sum_{i=1}^{n} X_i\right)^2$$

This number is called the *sample variance*. Interpret Exercise 6 in terms of the sample variance and the variance of the probability distribution of the identically distributed random variables $X_1, X_2, \ldots, X_n$.

8. In n Bernoulli trials with common probability p of success in each trial, let X_i be the number of successes in the ith trial, $i = 1, 2, \ldots, n$. Show that the sample variance of $X_1, X_2, \ldots, X_n$ is

$$\frac{S_n}{n} - \left(\frac{S_n}{n}\right)^2 = \frac{S_n}{n}\left(1 - \frac{S_n}{n}\right)$$

where S_n is the total number of successes.

9. In Exercise 8, find the expectation and the variance of $(S_n/n)(1 - S_n/n)$. On the basis of these findings, can you show directly that

$$P^n\left(\left|\frac{S_n}{n}\left(1 - \frac{S_n}{n}\right) - p(1-p)\right| > \epsilon\right) \to 0 \quad \text{as} \quad n \to \infty$$

for each $\epsilon > 0$?

10. For each n, let $X_1, X_2, \ldots, X_n$ be independent and identically distributed with some fixed probability distribution having a positive variance σ^2. Show that for each $\epsilon > 0$,

$$P^n\left(\left|\sqrt{\frac{1}{n}\sum_{i=1}^{n} X_i^2 - \left(\frac{1}{n}\sum_{i=1}^{n} X_i\right)^2} - \sigma\right| > \epsilon\right) \to 0 \quad \text{as} \quad n \to \infty$$

[HINT: Write

$$\left|\sqrt{\frac{1}{n}\sum_{i=1}^{n} X_i^2 - \left(\frac{1}{n}\sum_{i=1}^{n} X_i\right)^2} - \sigma\right| = \frac{\left|\frac{1}{n}\sum_{i=1}^{n} X_i^2 - \left(\frac{1}{n}\sum_{i=1}^{n} X_i\right)^2 - \sigma^2\right|}{\left|\sqrt{\frac{1}{n}\sum_{i=1}^{n} X_i^2 - \left(\frac{1}{n}\sum_{i=1}^{n} X_i\right)^2} + \sigma\right|}$$

and note that $\left|\sqrt{\frac{1}{n}\sum_{i=1}^{n} X_i^2 - \left(\frac{1}{n}\sum_{i=1}^{n} X_i\right)^2} + \sigma\right|$ is always larger than some positive number m. Hence the event that

$$\left| \sqrt{\frac{1}{n}\sum_{i=1}^{n} X_i^2 - \left(\frac{1}{n}\sum_{i=1}^{n} X_i\right)^2} - \sigma \right| > \epsilon$$

is contained in the event that

$$\left| \frac{1}{n}\sum_{i=1}^{n} X_i^2 - \left(\frac{1}{n}\sum_{i=1}^{n} X_i\right)^2 - \sigma^2 \right| > \epsilon m$$

Apply Exercise 6.]

11. For each n, let $X_1, X_2, \ldots, X_n$ be independent and identically distributed with some fixed probability having kth moment about the origin μ_k, $k > 2$. Show that for each $\epsilon > 0$,

$$P^n\left(\left|\frac{1}{n}\sum_{i=1}^{n} X_i^k - \mu_k\right| > \epsilon\right) \to 0 \quad \text{as} \quad n \to \infty$$

*8.4 The normal approximation to the binomial distribution

Let S_n be the number of successes in n Bernoulli trials with p the probability of success in each trial, $0 < p < 1$. As was seen in Chapter Six, S_n is a binomial random variable with parameters n and p. Hence, the probability that $S_n = r$ is given by $\binom{n}{r} p^r q^{n-r}$, $r = 0, 1, 2, \ldots, n$.

There are many problems in which we are not as interested in the probability that $S_n = r$ as we are in the probability that S_n takes some value between two integers, say α and β. As one important example of this, notice that the event "the relative frequency S_n/n of successes in n trials will differ from the probability p of a success by no more than $\frac{1}{10}$" (that is, $-\frac{1}{10} \leq S_n/n - p \leq \frac{1}{10}$) is the same as the event $n(p - \frac{1}{10}) \leq S_n \leq n(p + \frac{1}{10})$. (In this example, the bounds for S_n may not be integers. However, they could be replaced by appropriate integers, since S_n takes only integer values.)

First, let us inspect $P^n(\alpha \leq S_n \leq \beta)$, the probability that S_n takes some value between α and β inclusive. It is easy to write out this probability. Indeed,

$$P^n(\alpha \leq S_n \leq \beta) = \sum_{r=\alpha}^{\beta} \binom{n}{r} p^r q^{n-r} \tag{8.11}$$

However, if n and $\beta - \alpha$ are even moderately large, a direct computation of the sum in Eq. (8.11) may be prohibitively tedious. For example, if we want to compute the probability that in 500 tosses of a balanced coin the number of heads is between 200 and 300, we would have to compute

$$\sum_{r=200}^{300} \binom{500}{r} \frac{1}{2^{500}}$$

which is indeed a tedious task. However, when n is large, it is possible to approximate expressions like (8.11) without undue labor. This approximation technique was first discovered by DeMoivre (1667–1754) and Laplace (1749–1827), and the theorem embodying this discovery is called the *DeMoivre-Laplace limit theorem.* Historically, it is the first deep limit theorem of probability theory. It is best stated in terms of the standardized random variable S_n^* defined by $S_n^* = (S_n - np)/\sqrt{npq}$ (recall that standardized random variables have expectations 0 and variance 1). It discusses the limiting behavior of the probabilities $P^n(a \le S_n^* \le b)$ as $n \to \infty$, where a and b are given real numbers.

Theorem 8.3 *DeMoivre-Laplace Limit Theorem* Let a and b be any real numbers, $a < b$. Then,

$$P^n(a \le S_n^* \le b) \to \Phi(b) - \Phi(a) \quad \text{as} \quad n \to \infty \tag{8.12}$$

where, for any real number r, $\Phi(r)$ is defined by

$$\Phi(r) = \frac{1}{\sqrt{2\pi}} \int_{-\infty}^{r} e^{-x^2/2}\, dx \tag{8.13}$$

The proof of this theorem, although it requires only elementary techniques, is somewhat long. Therefore, we shall not reproduce it here. If you wish to study the proof, refer to one in the references quoted at the end of this book. Here, we only discuss the significance of this theorem and attempt to display its usefulness in solving problems.

The theorem says that, for large n, the probability that $a \le S_n^* \le b$ is close to $\Phi(b) - \Phi(a)$. Now $\Phi(b)$ and $\Phi(a)$ are, in principle, known. They are, respectively,

$$\frac{1}{\sqrt{2\pi}} \int_{-\infty}^{b} e^{-x^2/2}\, dx \quad \text{and} \quad \frac{1}{\sqrt{2\pi}} \int_{-\infty}^{a} e^{-x^2/2}\, dx$$

If somehow we could compute these integrals in terms of a and b, we would be able to find $\Phi(b) - \Phi(a)$. Then we would be able to use this value of $\Phi(b) - \Phi(a)$ as an approximation for $P^n(a \le S_n^* \le b)$ when n is large.

The function $f(x) = (1/\sqrt{2\pi})e^{-x^2/2}$, called the *normal density function,* is of great theoretical importance. It appears in the statement of a wide variety of limit theorems in probability theory, as it does in the DeMoivre-Laplace theorem. Its central role in probability theory has been satisfactorily explained only within the last 40 years or so, and its appearance in a wide variety of seemingly dissimilar contexts was regarded until then with some mystification. The function $f(x)$ was studied first by DeMoivre and Laplace. It was also extensively studied later on by Gauss (1777–1855) and is sometimes called the Gaussian function. The function $\Phi(r) = \int_{-\infty}^{r} f(x)\,dx$ is similarly an important function and is called the *normal distribution function.* These two functions are briefly discussed in the appendix.

There is no formula for the function $\Phi(r)$ in closed form. However, tables are available from which values of $\Phi(r)$ for various values of r can be computed. For example, see Table A.1 of the appendix where it is shown how $\Phi(r)$ can be found for various values of r.

Using the tabulated values of $\Phi(r)$, it is a simple matter to compute $\Phi(b) - \Phi(a)$ and thus to get an approximate value of $P^n(a \le S_n^* \le b)$ when n is large. On the other hand, the event $a \le S_n^* \le b$ is the same as the event $a\sqrt{npq} \le S_n - np \le b\sqrt{npq}$, which is the same as the event $np + a\sqrt{npq} \le S_n \le np + b\sqrt{npq}$. Thus, for large n, we have

$$P^n(np + a\sqrt{npq} \le S_n \le np + b\sqrt{npq}) \cong \Phi(b) - \Phi(a) \tag{8.14}$$

Letting $\alpha = np + a\sqrt{npq}$, $\beta = np + b\sqrt{npq}$ so that

$$a = \frac{\alpha - np}{\sqrt{npq}} \quad \text{and} \quad b = \frac{\beta - np}{\sqrt{npq}}$$

we find that for large n, Eq. (8.14) becomes

$$P^n(\alpha \le S_n \le \beta) \cong \Phi\left(\frac{\beta - np}{\sqrt{npq}}\right) - \Phi\left(\frac{\alpha - np}{\sqrt{npq}}\right) \tag{8.15}$$

Equation (8.15) is referred to as the normal approximation to the binomial distribution. It is usually an excellent approximation if n is sizable, say $n \ge 100$. If p is not too close to zero or to one, then this approximation is a surprisingly good one for values of n as small as 20.

For small values of n, a better approximation than Eq. (8.15) is given by

$$P^n(\alpha \le S_n \le \beta) \cong \Phi\left(\frac{(\beta - np + \frac{1}{2})}{\sqrt{npq}}\right) - \Phi\left(\frac{\alpha - np - \frac{1}{2}}{\sqrt{npq}}\right) \tag{8.16}$$

As n gets larger, the difference between the two approximations (8.15) and (8.16) becomes negligible.

It is possible on the basis of Eq. (8.15) to approximate the probability that $S_n \leq \beta$ also. In particular, the event $S_n \leq \beta$ is the same as the event $0 \leq S_n \leq \beta$, so

$$P^n(S_n \leq \beta) = P^n(0 \leq S_n \leq \beta)$$

$$\cong \Phi\left(\frac{\beta - np}{\sqrt{npq}}\right) - \Phi\left(\frac{-np}{\sqrt{npq}}\right) \tag{8.17}$$

Now $\Phi(-np/\sqrt{npq}) = \Phi(-\sqrt{np}/\sqrt{q})$, and for large n, this is approximately zero (see Table A.1 of the appendix). Therefore, for large n,

$$P^n(S_n \leq \beta) \cong \Phi\left(\frac{\beta - np}{\sqrt{npq}}\right) \tag{8.18}$$

We shall illustrate these approximations by two examples before we go on to discuss some important problems in which they can be utilized.

Example 8.7

Suppose $p = \frac{1}{2}$ and $n = 200$, and let us find approximately the value of $P^n(90 \leq S_n \leq 110)$. The relevance to coin tossing games is obvious. Here $np = 100$ and $npq = 50$. Setting $\alpha = 90$ and $\beta = 110$, we see that $(\alpha - np)/\sqrt{npq} = -10/\sqrt{50} = -1.414$, and $(\beta - np)/\sqrt{npq} = 10/\sqrt{50} = 1.414$. Thus, using Eq. (8.15), we get

$$P^n(90 \leq S_n \leq 110) \cong \Phi(1.414) - \Phi(-1.414) \tag{8.19}$$

Evaluating the right side from the tables (Table A.1 of the appendix), we see that the value of $P^n(90 \leq S_n \leq 100)$ is approximately .8426. The actual value (which can be computed with much labor by using tables of binomial probabilities) is .8680. The approximation is in error by .0154, i.e., by less than 2%. If, instead of using Eq. (8.15), we had used Eq. (8.16), we would have obtained $(\alpha - np - \frac{1}{2})/\sqrt{npq} = -10.5/\sqrt{50} = -1.485$ and $(\beta - np + \frac{1}{2})/\sqrt{npq} = +1.485$, and so we would have had

$$P^n(90 \leq S_n \leq 110) \cong \Phi(1.485) - \Phi(-1.485) \tag{8.20}$$

From the tables we find that this is equal to .8625, which is very near the true value.

Example 8.8

Let $p = .1$ and $n = 500$, and let us find $P^n(50 \leq S_n \leq 55)$ approximately. Here $np = 50$ and $npq = 45$. Letting $\alpha = 50$ and $\beta = 55$, we see that $(\alpha - np)/\sqrt{npq} = 0$ and $(\beta - np)/\sqrt{npq} = 5/\sqrt{45} = .7455$.

Applying Eq. (8.15), we get

$$P^n(50 \le S_n \le 55) \cong \Phi(.7455) - \Phi(0) = .2720 \tag{8.21}$$

The true value of $P^n(50 \le S_n \le 55)$ in this case is .3175; thus, the result in Eq. (8.21) is in error by 12%. If the finer approximation (8.16) had been used here, we would have had $(\alpha - np - \frac{1}{2})/\sqrt{npq} = -.5/\sqrt{45} = -.0746$, and $(\beta - np + \frac{1}{2})/\sqrt{npq} = 5.5/\sqrt{45} = .8201$. Thus,

$$P^n(50 \le S_n \le 55) \cong \Phi(.8201) - \Phi(-.0746) = .3235 \tag{8.22}$$

This approximation is in error by about 2%.

We shall conclude this section with applications of the DeMoivre-Laplace limit theorem to three problems that typically arise in statistical work. Note, however, that we are merely scratching the surface of a complicated subject, and our discussion is intended only to suggest a large variety of problems.

Example 8.9

A Confidence Interval for an Unknown p Suppose a market research organization wishes to estimate the proportion of adult males in a large community who use a particular brand of shaving cream. Suppose further that this is to be done by questioning those selected in a random sample of 200 adult males drawn with replacement. If p is the unknown proportion, the number S_{200} of successes (i.e., users) in the sample is a binomial random variable with parameters 200 and p. For short, we will omit the subscript in S_{200} and call it S. It is to be expected that when such a sampling is done, the difference $S/200 - p$ is likely to be small.

To be more precise, we may write, using the normal approximation to the binomial distribution,

$$P^{200}\left(a \le \frac{S - 200p}{\sqrt{200pq}} \le b\right) \cong \Phi(b) - \Phi(a) \tag{8.23}$$

In particular, let us take $a = -2$ and $b = 2$, so that

$$P^{200}\left(-2 \le \frac{S - 200p}{\sqrt{200pq}} \le 2\right) \cong \Phi(2) - \Phi(-2) \cong .955 \tag{8.24}$$

Thus, with an approximate probability of .955, the ratio $(S - 200p)/\sqrt{200pq}$ falls between -2 and 2 or, what is the same thing,

$$\left(\frac{S - 200p}{\sqrt{200pq}}\right)^2 \le 4 \tag{8.25}$$

Now, for each fixed number of successes S, Eq. (8.25) can only be satisfied for certain range of value of p. In fact, Eq. (8.25) holds only for those values of p for which

$$S^2 - 400Sp + 40000p^2 \leq 4(200p(1 - p)) \tag{8.26}$$

or for which

$$40800p^2 - (400S + 800)p + S^2 \leq 0 \tag{8.27}$$

Recalling the roots of a quadratic equation, we see that Eq. (8.27) can only hold for values of p satisfying

$$\frac{S}{204} + \frac{1}{102} - \frac{\sqrt{400 + 400S - 2S^2}}{2040} \leq p \leq \frac{S}{204} + \frac{1}{102} + \frac{\sqrt{400 + 400S - 2S^2}}{2040} \tag{8.28}$$

Equation (8.28) is interpreted as follows. With an approximate probability of .955, the random variable

$$\frac{S}{204} + \frac{1}{102} - \frac{\sqrt{400 + 400S - 2S^2}}{2040}$$

will be at least as small as the unknown proportion p at the same time that the random variable

$$\frac{S}{204} + \frac{1}{102} + \frac{\sqrt{400 + 400S - 2S^2}}{2040}$$

is at least as large as p. It is then customary to compute these two random variables and state that p lies between them. This statement may be incorrect, but the probability of making a correct statement using this procedure is approximately .955. The interval between the two random variables is called a *confidence interval* for p and in this case would be said to have a *confidence level* of approximately .955.

For example, if the sampling produces a value for S of 40,

$$\frac{S}{204} + \frac{1}{102} - \frac{\sqrt{400 + 400S - 2S^2}}{2040} \cong \frac{61}{408}$$

and

$$\frac{S}{204} + \frac{1}{102} + \frac{\sqrt{400 + 400S - 2S^2}}{2040} \cong \frac{107}{408}$$

We would then state that the interval from .15 to .262 contains p, that is, $.15 \leq p \leq .262$. Again note that this statement may be incorrect and p might be .3, for example. However, if the above procedure were followed many times, then the confidence interval obtained would turn out to straddle the true value of p, approximately 95.5% of the time.

This same kind of analysis may be done for other choices besides $a = -2$ and $b = 2$. For example, if we take $a = -1.64$ and $b = 1.64$, two changes would be evident in the above discussion. First, the confidence *level* would now be smaller, viz., it is approximately $\Phi(1.64) - \Phi(-1.64)$ or, about .90. Second, the confidence *intervals* would be somewhat shorter. Overall then, the bounds placed on p would be closer bounds, but this is paid for to the extent that confidence in these bounds is not as high.

Example 8.10

A Determination of Sample Size Rather than deciding somewhat arbitrarily on a random sample of size 200 as we did in Example 8.9, the market research organization might ask for a certain specified accuracy in the estimate of p and then take a sample large enough to achieve this. Let us take the accuracy requirement to be that S_n/n should differ from the unknown proportion p by no more than .01. Now this accuracy cannot be achieved with certainty no matter how large the sample size n may be, for we can imagine extreme situations, such as always drawing successes or always drawing failures, so that our estimate for p can be either zero or one. If we modify the requirements to take this uncertainty into account, it could be asked that the event "S_n/n differs from p by no more than .01" have a large probability, say .95. The question then is whether we can ensure this by taking a large enough sample size n.

We wish to find n so large that

$$P^n\left(\left|\frac{S_n}{n} - p\right| \leq .01\right) \geq .95 \tag{8.29}$$

Now

$$\frac{S_n}{n} - p = \frac{\sqrt{pq}}{\sqrt{n}}\left(\frac{S_n - np}{\sqrt{npq}}\right)$$

so by the normal approximation to the binomial distribution,

$$\begin{aligned} P^n\left(-.01 \leq \frac{S_n}{n} - p \leq .01\right) \\ &= P^n\left(-.01 \leq \frac{\sqrt{pq}}{\sqrt{n}}\left(\frac{S_n - np}{\sqrt{npq}}\right) \leq .01\right) \\ &= P^n\left(-.01\frac{\sqrt{n}}{\sqrt{pq}} \leq \left(\frac{S_n - np}{\sqrt{npq}}\right) \leq .01\frac{\sqrt{n}}{\sqrt{pq}}\right) \\ &\cong \Phi\left(.01\frac{\sqrt{n}}{\sqrt{pq}}\right) - \Phi\left(-.01\frac{\sqrt{n}}{\sqrt{pq}}\right) \end{aligned} \tag{8.30}$$

Thus, we want n so large that

$$\Phi\left(.01\frac{\sqrt{n}}{\sqrt{pq}}\right) - \Phi\left(-.01\frac{\sqrt{n}}{\sqrt{pq}}\right) \geq .95 \tag{8.31}$$

Since $\Phi(1.96) - \Phi(-1.96) = .95$, this may be accomplished if

$$.01\frac{\sqrt{n}}{\sqrt{pq}} \geq 1.96 \quad \text{or} \quad n \geq (196)^2 pq \tag{8.32}$$

We find in Eq. (8.32) that the required sample size depends on the value of the proportion p, which is unknown. This difficulty may be circumvented by taking a conservative approach. Namely, let us take n to be large enough to satisfy $n \geq (196)^2 pq$ for *any* choice of p. It may be verified that $pq = p(1 - p)$ is never any larger than $\frac{1}{4}$. If we choose n so large that n satisfies $n \geq (196)^2 \frac{1}{4}$, then we can be sure that $n \geq (196)^2 pq$ no matter what p is. Therefore, a sample size of size at least $(196)^2 \frac{1}{4} = 9604$ will ensure that with probability .95, S_n/n will differ from the true value of p by no more than .01.

It is possible to set other accuracy requirements than that S_n/n differ from p by no more than .01 with probability .95. Generally, of course, the greater the accuracy and confidence required, the larger will be the sample size necessary to achieve it.

Example 8.11

Deciding Whether or Not p Is Some Specified Value In attempting to determine whether or not a subject may be said to have extrasensory perception, something akin to the following procedure is commonly used. The subject is placed in one room, while, in another room, a card is randomly selected from an ordinary deck. The subject is then asked to guess the suit of the card drawn, i.e., he is to guess spades, hearts, diamonds, or clubs.

In this experiment, a person with no extrasensory powers might be expected to guess correctly with probability $\frac{1}{4}$. In n trials, such a person will get approximately $n/4$ correct answers (successes). On the other hand, a person who claims that he has extrasensory powers should presumably be able to guess correctly in a substantially larger proportion of trials.

Let us say, for the sake of being definite, that 300 such trials are performed with a given person guessing, and let us suppose S_{300} (which we shall call S for short) of these result in correct guesses. We now ask the question: how large should the value of S be in order for us to lend credence to the subject's claim that he has extrasensory powers? This is indeed a very complicated question. For, even though the subject may be merely guessing, it is conceivable that on a particular occasion a large proportion of guesses turns out to be correct, without this being a recurrent feature of the subject's response to the experiment. Rather than give a direct answer to the question, we may argue as follows. *If* the subject were merely guessing, we could regard the 300 trials as Bernoulli trials with a probability $p = \frac{1}{4}$ in a given trial. Then, from Eq. (8.18),

$$P^{300}(S \leq \beta) \cong \Phi(\tfrac{2}{15}(\beta - 75)) \tag{8.33}$$

Three special cases of Eq. (8.33) are

$$\begin{aligned} P^{300}(S \leq 85) &\cong \Phi(\tfrac{2}{15}(85 - 75)) = \Phi(\tfrac{4}{3}) \cong .908 \\ P^{300}(S \leq 90) &\cong \Phi(\tfrac{2}{15}(90 - 75)) = \Phi(2) \cong .977 \\ P^{300}(S \leq 95) &\cong \Phi(\tfrac{2}{15}(95 - 75)) = \Phi(\tfrac{8}{3}) \cong .996 \end{aligned} \tag{8.34}$$

Thus, for example, for a *normal* subject, the probability that the number of successes S is less than 85 is about .908. From the standpoint of frequencies, if 1000 *normal* subjects were each given 300 trials, we could expect about 908 of them to score no more than 85 correct, about 955 to score no more than 90 correct, and about 996 to score no more than 95 correct.

Now suppose that for some unusually gifted subject p is not $\frac{1}{4}$ but rather .4. Then we would have

$$P^{300}(S \leq \beta) \cong \Phi\left(\frac{\beta - 120}{\sqrt{72}}\right) \tag{8.35}$$

Three special cases of Eq. (8.35) are

$$\begin{aligned} P^{300}(S \leq 85) &\cong \Phi\left(\frac{85 - 120}{\sqrt{72}}\right) \cong \Phi(-4.12) \cong 0 \\ P^{300}(S \leq 90) &\cong \Phi\left(\frac{90 - 120}{\sqrt{72}}\right) \cong \Phi(-3.54) \cong .0002 \\ P^{300}(S \leq 95) &\cong \Phi\left(\frac{95 - 120}{\sqrt{72}}\right) \cong \Phi(-2.95) \cong .0016 \end{aligned} \tag{8.36}$$

Equations (8.36) could be interpreted in the same way as was Eq. (8.35).

If we were obligated to classify a large number of subjects as being either normal or extrasensory on the basis of 300 trials, we might adopt one of the three strategies:

$$\text{Classify}\quad \begin{aligned} &(1)\ \text{subject normal if the score is} \leq 85 \\ &\text{extrasensory if the score is} > 85 \\ &(2)\ \text{subject normal if the score is} \leq 90 \\ &\text{extrasensory if the score is} > 90 \\ &(3)\ \text{subject normal if the score is} \leq 95 \\ &\text{extrasensory if the score is} > 95 \end{aligned} \tag{8.37}$$

The first strategy would incorrectly classify about 9.2% of all normal subjects as extrasensory, the second about 4.5%, and the third about .4%. Of all subjects extrasensory to the extent of having $p = .4$, less than .005% would be incorrectly classified as normal by the first strategy, about .02% would be incorrectly classified by the second, and about .84% by the third.

The above considerations only suggest the complexity of the mathematical situation. The practical question of extrasensory perception is more complex than this.

EXERCISES 8.4

1. Suppose a day in some even climate has probability $\frac{2}{3}$ of being sunny and probability $\frac{1}{3}$ of being cloudy. What is the approximate probability that the number of sunny days in a year is no less than 200 and no more than 250? (Assume Bernoulli trials.) What is the approximate probability that there are at least 200 sunny days?

2. A student entering a university is assumed to choose a humanities major with probability $\frac{3}{5}$ and a science major with probability $\frac{2}{5}$. What is the approximate probability that in an entering freshman class of 500 students between 175 and 225 students will choose a science major? (Assume Bernoulli trials.)

3. Suppose items coming off a production line are defective with probability $\frac{1}{20}$. If these items are boxed, 1000 items to a box, what is the approximate probability that a box contains no more than 60 defective items? (Assume Bernoulli trials.)

4. A balanced coin is flipped 400 times. Determine the number x such that the probability that the number of heads is between $200 - x$ and $200 + x$ is approximately .85. (Assume Bernoulli trials.)

5. In a large community, $\frac{2}{3}$ of the adult males own cars. If a random sample of size 300 is drawn with replacement from among adult males, find the number x such that the approximate probability that the number of car owners in the sample is between $200 - x$ and $200 + x$ is .90.

6. A salesman finds that he makes a sale on a call with probability $\frac{1}{10}$. How many calls should he make in a year so that the probability that he makes less than 80 sales is no more than .05? (Assume Bernoulli trials.)

7. How large must n be in order that $\Phi(-\sqrt{np}/\sqrt{1-p})$ be less than .01 when $p = \frac{1}{2}$? When $p = \frac{1}{4}$? When $p = \frac{1}{8}$? [See Eq. (8.17) and the remark following it.]

8. Give an approximation to $P^n(S_n \geq \alpha)$.

9. Find a confidence interval for an unknown p based on 100 Bernoulli trials which has an approximate confidence level of .90. What does this confidence interval become when the number of successes is 50? When the number of successes is 60?

10. How large a sample size n is required in Bernoulli trials to bring S_n/n within .02 of p with approximate probability .95? To bring S_n/n within .001 of p with approximate probability .95?

11. How large a sample size n is required in Bernoulli trials so that S_n/n is at least $p - .01$ with approximate probability .95?

12. Using the three strategies listed in Eq. (8.37), find approximately the probability that a hypothetical "extrasensory" subject for whom $p = \frac{1}{3}$ would be misclassified as normal by the use of each of these strategies.

*8.5 The central limit theorem

We shall now state a more general limit theorem than the DeMoivre-Laplace limit theorem and sketch some applications of it, analogous to those discussed in Section 8.4.

We let $X_1, X_2, \ldots, X_n$ be mutually independent, identically distributed random variables with an arbitrary probability distribution, and we adopt the setting of Section 8.2. Also, we set

$$S_n = X_1 + X_2 + \cdots + X_n$$

As in the case of the DeMoivre-Laplace theorem, we would like to get some idea of the probability distribution of S_n. That is, given real numbers α and β, we would like to obtain an estimate for the probability of the event $\alpha \leq S_n \leq \beta$.

In the case of Bernoulli trials, the probability of this event was in principle exactly known, but it was cumbersome to compute. In the present case, in the absence of a specific form for the common probability distribution of $X_1, X_2, \ldots, X_n$, we cannot of course claim that the probability of $\alpha \leq S_n \leq \beta$ can be computed in principle. Nevertheless it is a remarkable fact that this probability can be approximated in a rather simple way provided n is large. This theorem is known as the *central limit theorem*. We shall content ourselves with stating it and commenting on it, without proving it. We shall also illustrate some of its commonest applications.

Since $X_1, X_2, \ldots, X_n$ have the same distribution, they have a common expectation, say μ, and a common variance, say σ^2. It is then clear that $E(S_n) = n\mu$ and $V(S_n) = n\sigma^2$. Therefore, the standard deviation of S_n is $\sigma\sqrt{n}$. We now introduce the so-called standardized random variable S_n^*, defined by $S_n^* = (S_n - n\mu)/\sigma\sqrt{n}$. Note that S_n^* can equivalently be defined by

$$S_n^* = \frac{S_n/n - \mu}{\sigma/\sqrt{n}}$$

The central limit theorem can be stated very simply in terms of S_n^*. It is a result about the behavior of the probability of an event of the type $a \leq S_n^* \leq b$.

Theorem 8.4 *The Central Limit Theorem* For each n, let $X_1, X_2, \ldots, X_n$ be mutually independent identically distributed random variables on (Ω^n, P^n). Let μ be their common expectation and $\sigma^2 > 0$ be their common variance. Let $S_n = X_1 + X_2 + \cdots + X_n$ and $S_n^* = (S_n - n\mu)/\sigma\sqrt{n}$. If a and b are any real numbers, say $a \leq b$, then

$$P^n(a \leq S_n^* \leq b) \rightarrow \Phi(b) - \Phi(a) \quad \text{as} \quad n \rightarrow \infty \tag{8.38}$$

Here $\Phi(r)$ is the function defined by

$$\Phi(r) = \frac{1}{\sqrt{2\pi}} \int_{-\infty}^{r} e^{-x^2/2}\, dx \tag{8.39}$$

Although the proof of this theorem is beyond the scope of this book, its statement is very simple, and our work in earlier chapters puts some of the applications of this theorem within our ken.

Note the close similarity of the central limit theorem with the DeMoivre-Laplace theorem. Indeed, the latter is but a special case of the former. Historically, the central limit theorem is fairly recent, even in the special case we are discussing, dating back less than 50 years. It is valid for many random variables a good deal more general than those defined on finite sample spaces. As a limit theorem of probability theory, it is indeed one of the most distinguished.

The remarkable facet of the central limit theorem is the fact that the limiting behavior described by Eq. (8.38) holds *irrespective of what the common probability distribution of the random variables* $X_1, X_2, \ldots, X_n$ *happens to be.* It appears therefore that the function $\Phi(r)$ attracts hordes of distributions to it in limiting form.

Like the DeMoivre-Laplace theorem, the central limit theorem enables us to find approximately the value of $P^n(a \leq S_n^* \leq b)$, if n is large, by finding $\Phi(b)$ and $\Phi(a)$ from the tabulated values of the function Φ. Moreover, given real numbers α and β, we quickly see that the event $\alpha \leq S_n \leq \beta$ is the same as the event $a \leq S_n^* \leq b$ where $a = (\alpha - n\mu)/\sigma\sqrt{n}$ and $b = (\beta - n\mu)/\sigma\sqrt{n}$. Thus, we have, for large n,

$$P^n(\alpha \leq S_n \leq \beta) \cong \Phi\left(\frac{\beta - n\mu}{\sigma\sqrt{n}}\right) - \Phi\left(\frac{\alpha - n\mu}{\sigma\sqrt{n}}\right) \tag{8.40}$$

Since σ and μ are parameters determined by the common distribution of $X_1, X_2, \ldots, X_n$, they are in principle known and Eq. (8.40) enables us to compute approximately the probability $P^n(a \leq S_n \leq b)$.

Finally, given any real numbers c and d, we may check that the event $c \leq S_n/n \leq d$ is the same as the event $nc \leq S_n \leq nd$, which in turn is the same as the event $a \leq S_n^* \leq b$, with

$$a = \frac{nc - n\mu}{\sigma\sqrt{n}} = \frac{c - \mu}{\sigma/\sqrt{n}} \quad \text{and} \quad b = \frac{nc - n\mu}{\sigma\sqrt{n}} = \frac{d - \mu}{\sigma/\sqrt{n}}$$

Hence, for large n, we have

$$P^n\left(c \leq \frac{S_n}{n} \leq d\right) \cong \Phi\left(\frac{d - \mu}{\sigma/\sqrt{n}}\right) - \Phi\left(\frac{c - \mu}{\sigma/\sqrt{n}}\right) \tag{8.41}$$

which is very frequently a useful form in which to remember Eq. (8.40).

Example 8.12

Suppose in a fund drive for a symphony orchestra a telephone call can produce the following result: no contribution with probability $\frac{3}{8}$; a regular \$10 membership with probability $\frac{1}{4}$; a \$25 sustaining membership with probability $\frac{1}{4}$; and a \$50 life membership with probability $\frac{1}{8}$. Suppose 1000 telephone calls may be made independently under these conditions. (Our assumptions are reasonable only when calls are made without replacement to a population much larger than 1000.) With what approximate probability will the total funds from these calls exceed \$14,000?

We easily see that the expectation from a single call is 15 while the variance is 1075/4. We will compute approximately the probability of the complementary event to the one mentioned above. Thus, we ask for the probability that $S_{1000} \leq 14{,}000$ or, what is the same thing, $0 \leq S_{1000} \leq 14{,}000$. Therefore, by Eq. (8.40),

$$\begin{aligned} P^{1000}(0 \leq S_{1000} \leq 14{,}000) &\cong \Phi\left(\frac{14{,}000 - 15{,}000}{\frac{\sqrt{1075}}{\sqrt{4}}\sqrt{1000}}\right) \\ &\quad - \Phi\left(\frac{0 - 15{,}000}{\frac{\sqrt{1075}}{\sqrt{4}}\sqrt{1000}}\right) \\ &= \Phi\left(-\frac{1000}{25\sqrt{430}}\right) - \Phi\left(-\frac{15{,}000}{25\sqrt{430}}\right) \\ &\cong \Phi(-1.93) - \Phi(-29) \cong .027 \end{aligned} \tag{8.42}$$

But then the probability that S_{1000} exceeds 14,000 is approximately .973.

Example 8.13

Recall the setting of Example 8.6 where we considered sampling from the age distribution of the population of the United States. Suppose again that there are N people in the population, of whom N_j are of age j, $j = 0, 1, 2, \ldots, 125$. Random sampling with replacement will produce random variables $X_1, X_2, \ldots, X_n$ which are independent and identically distributed with a common probability distribution given by $f_X(j) = N_j/N$, $j = 0, 1, 2, \ldots, 125$. The expectation of each X_i is $\mu = \sum_{j=0}^{125} j(N_j/N)$, which is also called the population average. The variance of each X_i is

$$\sigma^2 = \sum_{j=0}^{125} (j - \mu)^2 \frac{N_j}{N}$$

which is also called the population variance. We set

$$S_n = X_1 + X_2 + \cdots + X_n$$

and we write the sample average

$$\frac{S_n}{n} = \frac{X_1 + X_2 + \cdots + X_n}{n}$$

also as $\bar{X}$.

We have previously asserted that a random sample of size $n = 40{,}000$ would suffice to give a sample average $\bar{X}$ that deviates from the unknown population average μ by more than one with probability at most .01. This statement was based on the Chebychev inequality and a guess that $\sigma^2 \leq 400$, i.e., a guess that the standard deviation of age within the population is no more than 20 years.

If we continue to suppose a priori that $\sigma^2 \leq 400$, we may refine the above assertion in various ways on the basis of the central limit theorem. For example, we will assert the same accuracy holds for a smaller sample size than 40,000. This improvement may be traced to the fact that the Chebychev inequality merely bounds some pertinent probability, while the central limit theorem approximates it.

Let us ask how large should n be in order that X differ by no more than one from μ with probability at least .99. That is, we seek an n so large that

$$P^n(|\bar{X} - \mu| \leq 1) \geq .99 \tag{8.43}$$

Now $\bar{X} - \mu = \dfrac{\sigma}{\sqrt{n}}\left(\dfrac{\bar{X} - \mu}{\sigma/\sqrt{n}}\right) = \dfrac{\sigma}{\sqrt{n}} S_n^*$, so we ask that

$$P^n\left(\frac{\sigma}{\sqrt{n}}\left|\frac{\bar{X} - \mu}{\sigma/\sqrt{n}}\right| \leq 1\right) = P^n\left(\left|\frac{\bar{X} - \mu}{\sigma/\sqrt{n}}\right| \leq \frac{\sqrt{n}}{\sigma}\right)$$

$$= P^n\left(-\frac{\sqrt{n}}{\sigma} \leq \frac{\bar{X} - \mu}{\sigma/\sqrt{n}} \leq \frac{\sqrt{n}}{\sigma}\right) \tag{8.44}$$

$$\cong \Phi\left(\frac{\sqrt{n}}{\sigma}\right) - \Phi\left(-\frac{\sqrt{n}}{\sigma}\right) \geq .99$$

We find from Table 1 of the Appendix that $\Phi(2.57) - \Phi(-2.57) \cong .99$, so n should be large enough so that

$$\frac{\sqrt{n}}{\sigma} \geq 2.57 \quad \text{or} \quad n \geq (2.57)^2\sigma^2 \tag{8.45}$$

Inasmuch as we are assuming $\sigma^2 \leq 400$, then n will ensure the accuracy required provided $n \geq (2.57)^2 400 \cong 2640$. Because of the more refined estimate of probability available, we see that a much smaller sample size than 40,000 would be sufficient for the avowed purpose.

We assert, therefore, that a random sample of size 2640 drawn from the population of the United States with replacement will produce a sample average that differs from the population average by no more than 1 year with an approximate probability of .99.

Let us modify the problem of estimating μ as follows. Suppose a random sample of size 40,000 is to be taken. How small a deviation between the sample and population averages can be guaranteed with an approximate probability of .99? In other words, for how small a positive number ϵ is it true that

$$P^{40,000}(|\bar{X} - \mu| \leq \epsilon) \cong .99? \tag{8.46}$$

Transforming the event $|\bar{X} - \mu| \leq \epsilon$ to an event concerning $S_{40,000}^*$, we seek ϵ so that

$$P^{40,000}\left(\frac{\sigma}{200}\left|\frac{\bar{X} - \mu}{\sigma/200}\right| \leq \epsilon\right) = P^{40,000}\left(-\epsilon \leq \frac{\sigma}{200}\left(\frac{\bar{X} - \mu}{\sigma/200}\right) \leq \epsilon\right)$$

$$= P^{40,000}\left(\frac{-200\epsilon}{\sigma} \leq S_{40,000}^* \leq \frac{200\epsilon}{\sigma}\right) \tag{8.47}$$

$$\cong \Phi\left(\frac{200\epsilon}{\sigma}\right) - \Phi\left(\frac{-200\epsilon}{\sigma}\right) \cong .99$$

Equation (8.47) will be satisfied by a choice of ϵ for which $200\epsilon/\sigma \cong 2.57$ or for $\epsilon \cong 2.57\sigma/200$. This means that one could state with approximate probability .99 that $\bar{X}$ will differ from the true population average μ by less than $2.57\sigma/200$. This number can be evaluated as soon as we know σ. However, even if we do not know what σ is exactly, we may still know an upper bound for it. For example, suppose we are willing to assume, say, that $\sigma \leq 20$, then we may state that the probability of the event that $\bar{X}$ differs from μ by less than $(2.57)(20)/200$ $(= .257)$ is at least as large as the probability that $\bar{X}$ differs from μ by less than $2.57\sigma/200$. (Why?)

Therefore, a random sample of size 40,000 drawn with replacement from the population of the United States will produce a sample average age that differs from the population average age by no more than .257 with an approximate probability of at least .99.

In practice, one is not usually content with a guess about the population variance σ^2. In particular, our assertions could be made somewhat more precise if we knew σ^2. In the last two cases, if σ^2 were really 100, we would have found that a sample size of 660 would suffice to bring $\bar{X}$ within 1 year of μ with approximate probability .99 and we would have found that a sample of 40,000 would bring $\bar{X}$ within .1285 years of μ with approximate probability .99. A lack of knowledge about σ^2 would thus appear to be rather costly.

It is in fact possible to obtain an estimate of σ^2 from a random sample drawn with replacement from the population. Suppose $X_1, X_2, \ldots, X_n$ are the ages drawn in such a sample. In Exercise 7 of Section 8.3, we have defined the sample variance to be the random variable

$$\frac{1}{n}\sum_{i=1}^{n} X_i^2 - \left(\frac{1}{n}\sum_{i=1}^{n} X_i\right)^2 = \frac{1}{n}\sum_{i=1}^{n} X_i^2 - (\bar{X})^2 \tag{8.48}$$

Furthermore, in Exercise 6 of Section 8.3, it was pointed out that

$$P^n\left(\left|\frac{1}{n}\sum_{i=1}^{n} X_i^2 - (\bar{X})^2 - \sigma^2\right| > \epsilon\right) \to 0 \quad \text{as} \quad n \to \infty \tag{8.49}$$

for each positive number ϵ. In other words, the sample variance in a large random sample is likely to be close to the population variance σ^2.

One simple use of Eq. (8.49) is as follows. One random sample is drawn from the population and the sample variance is taken as an estimate of σ^2. A second random sample is drawn from the population for the purpose of estimating μ. In the latter sample, the previous sample variance is used *as if it is* σ^2.

There are other and more subtle uses of the sample variance as it is given in Eq. (8.48). However, we will not go into them here as they would carry us somewhat far afield.

EXERCISES 8.5

1. Suppose that a balanced die is thrown 2000 times and that these trials are independent. What is the approximate probability that the total score is no less than 7000 and no more than 8000? Within what limits of the form $7000 - x$ and $7000 + x$ will the total score be with an approximate probability of .5?

2. Suppose a random sample of size 20,000 is drawn from a standard deck of cards with replacement. What is the approximate probability that the total number of aces that are drawn is no less than 1400 and no more than 1600? Is no less than 1400?

3. Consider an experiment consisting of drawing a random sample of size five from a standard deck of cards without replacement. If 4000 independent trials of this experiment are made under identical conditions, what is the approximate probability that the total number of aces drawn is no less than 1400 and no more than 1600? Is no less than 1400? Explain the difference between the answers to this exercise and Exercise 2.

4. A bridge player observes that in the course of a year he has been dealt 1000 bridge hands and the total honor count in these hands has been 9000. Is he justified in complaining about the general lack of honor count? (Assume independent trials. The expectation and variance of the honor count in a single hand have been found in Chapter Seven. Compute the approximate probability that in 1000 hands the total honor count will not be more than 9000.)

5. An encyclopedia salesman finds that on a given call he makes no sale with probability $\frac{1}{2}$; he sells the regular edition at \$50 with probability $\frac{3}{8}$; and he sells the custom edition at \$100 with probability $\frac{1}{8}$. How many calls should he make within a year to insure total sales of \$5000 with an approximate probability of .95? (Assume independent trials.)

6. Consider the field bet outlined in Example 7.2. Suppose a man makes bets of \$1 on a total of 3600 throws of the dice. What is the approximate probability that he does not lose any money? (Assume independent trials.)

7. Do Exercise 6 for 7200 throws and for 36,000 throws.

8. Suppose that a random sample of size 1000 is drawn from a population and that the numerical quantity being studied has an unknown population average μ and a known population variance of $\sigma^2 = 16$. With what approximate probability will the sample average deviate by no more than $\frac{1}{8}$ from the population average?

9. In Exercise 8, determine a positive ϵ for which the approximate probability that the sample average differs from the population average by more than ϵ is .05.

10. The central limit theorem is stated with the assumption that the common variance σ^2 of the random variances $X_1, X_2, \ldots, X_n$ is positive. Explain why the theorem cannot be true if $\sigma^2 = 0$. In particular, what is the probability distribution of S_n when $\sigma^2 = 0$?

11. The function Φ has the property that $\Phi(x) \to 1$ as $x \to \infty$ and $\Phi(-x) \to 0$ as $x \to \infty$. Using this, show that the weak law of large numbers is a consequence of the central limit theorem.

CHAPTER NINE

Random Variables with a Countable Number of Values

9.1 Introduction

We begin this chapter by considering the Poisson approximation to the binomial distribution. Then we are led to study random variables that are capable of assuming a countably infinite number of values. Our treatment of the latter topic is necessarily less comprehensive than that given for random variables assuming finitely many values. In Section 9.3, corresponding notions of probability distribution, expectation, and variance are defined for the countably infinite case, and we examine some applications of these topics.

*9.2 The Poisson approximation to the binomial distribution

Suppose we have a sequence of n Bernoulli trials with probability p of success in each trial. We have seen that, if S_n is the total number of successes obtained in these trials, the probability that $S_n = k$ is given by

$$f_{S_n}(k) = \binom{n}{k} p^k(1 - p)^{n-k}, \qquad 0 \leq k \leq n \tag{9.1}$$

When the number of trials n is large and the probability of success p is small (i.e., near 0), it may be demonstrated that

$$f_{S_n}(k) \cong e^{-\lambda} \frac{\lambda^k}{k!} \quad \text{where} \quad \lambda = np \tag{9.2}$$

This approximation to the binomial distribution is due to Poisson (1781–1840).

Before demonstrating how this approximation is obtained, let us consider two simple examples in which we might use it.

Example 9.1

A golf tournament committee decides to award \$10,000 to the first (if any) player to score a hole-in-one on a specified hole during the tournament. The probability that a player can make a hole-in-one is taken to be $\frac{1}{5000}$, and there is to be, during the course of the tournament, 3000 attempts. If it is assumed that 3000 Bernoulli trials are made with probability of success $\frac{1}{5000}$ at each trial, then the probability that the prize is not awarded is exactly $(\frac{4999}{5000})^{3000}$. To apply the Poisson approximation, we set $\lambda = np = 3000 \cdot \frac{1}{5000} = .6$ and $k = 0$. The probability that the prize is not awarded is approximately $e^{-.6}$, which from a table of natural logarithms we find to be about .55.

Example 9.2

A man regularly buys a ticket in an office pool. Each week 500 tickets are sold, and the winning ticket is drawn at random from among them. If the man buys a ticket for each of 1500 weeks, what is the probability that he wins at most twice?

Viewing this process as 1500 Bernoulli trials with probability $\frac{1}{500}$ of success in each trial, we see that the probability is

$$\left(\frac{499}{500}\right)^{1500} + \binom{1500}{1}\left(\frac{1}{500}\right)\left(\frac{499}{500}\right)^{1499} + \binom{1500}{2}\left(\frac{1}{500}\right)^2\left(\frac{499}{500}\right)^{1498}$$

For the Poisson approximation in Eq. (9.2), we set

$$\lambda = np = 1500 \cdot \tfrac{1}{500} = 3$$

and successively $k = 0, 1, 2$. The approximate probability is then

$$e^{-3} + e^{-3}\frac{3^1}{1!} + e^{-3}\frac{3^2}{2!} = \frac{17}{2}e^{-3}$$

or about .42.

To show the approximation (9.2), we proceed as follows. Consider n Bernoulli trials with probability of success λ/n in each trial, λ being a fixed positive number. If S_n is the total number of successes in these n trials, we will investigate the behavior of $f_{S_n}(k)$ as $n \to \infty$ with k a fixed nonnegative integer. Notice especially that in this process the probability of success, when n trials are considered, depends on the value of n.

We have

$$
\begin{aligned}
f_{S_n}(k) &= \binom{n}{k}\left(\frac{\lambda}{n}\right)^k\left(1-\frac{\lambda}{n}\right)^{n-k} \\
&= \frac{n(n-1)\cdots(n-k+1)}{k!}\left(\frac{\lambda}{n}\right)^k\left(1-\frac{\lambda}{n}\right)^{n-k} \\
&= \frac{1}{k!}\frac{n(n-1)\cdots(n-k+1)}{n^k}\lambda^k\left(1-\frac{\lambda}{n}\right)^n\left(1-\frac{\lambda}{n}\right)^{-k} \\
&= \frac{\lambda^k}{k!}\left(\frac{n-1}{n}\right)\left(\frac{n-2}{n}\right)\cdots\left(\frac{n-k+1}{n}\right)\left(1-\frac{\lambda}{n}\right)^n\left(1-\frac{\lambda}{n}\right)^{-k}
\end{aligned}
\tag{9.3}
$$

In the last expression for $f_{S_n}(k)$, the behavior as $n \to \infty$ of each term that depends on n is easy to describe. Specifically, each term of the form $(n - i)/n$ tends to 1, $(1 - \lambda/n)^{-k}$ tends to 1, and it is a basic result of the calculus that $(1 - \lambda/n)^n$ tends to $e^{-\lambda}$. In other words, $f_{S_n}(k)$ tends to $e^{-\lambda}(\lambda^k/k!)$ as $n \to \infty$, and this is the basis of the Poisson approximation to the binomial distribution.

From what we have just seen, Eq. (9.2) holds for n large and $p = \lambda/n$ small. Precisely how large and small these two numbers should be depends on the degree of accuracy desired. As a rough guide, if $p \leq .1$ and $n \geq 100$, reasonably good approximation can be expected.

For future reference, we make one further observation about the approximation (9.2). Equation (9.2) is applied to $f_{S_n}(k)$ when the number of trials is large and the probability of success on a given trial is small. Now, the numbers $f_{S_n}(k)$ are probabilities and satisfy, for $0 \leq k \leq n$,

$$f_{S_n}(k) > 0 \quad \text{and} \quad \sum_{k=0}^{n} f_{S_n}(k) = 1 \tag{9.4}$$

The numbers $e^{-\lambda}\dfrac{\lambda^k}{k!}$ for $k \geq 0$ satisfy a similar property, viz.,

$$e^{-k}\frac{\lambda^k}{k!} > 0 \quad \text{and} \quad \sum_{k=0}^{\infty} e^{-\lambda}\frac{\lambda^k}{k!} = 1 \tag{9.5}$$

The second part of Eq. (9.5) follows from the expansion formula

$$e^{\lambda} = \sum_{k=0}^{\infty} \frac{\lambda^k}{k!} \tag{9.6}$$

($\sum_{k=0}^{\infty}$ stands, of course, for an infinite series.)

EXERCISES 9.2

1. In Example 9.1, suppose an insurance company, in exchange for a premium payment, agrees to pay the \$10,000 in the event of a hole-in-one. What premium should the company charge in order that the expected profit from the transaction be about \$500?

2. In Example 9.2, what is the approximate probability that the man holds exactly three winning tickets? At most three winning tickets? More than four winning tickets?

3. Using a table of logarithms, compute the probability of having no successes in n Bernoulli trials with probability p of success in each trial when
(a) $n = 10$, $p = \frac{1}{10}$.
(b) $n = 100$, $p = \frac{1}{100}$.
(c) $n = 1000$, $p = \frac{1}{1000}$.
Compare these answers with the value of e^{-1}.

4. Suppose S is the total number of successes in 200 Bernoulli trials with a probability of success $\frac{1}{100}$ in each trial. Using the Poisson approximation, find the smallest number of successes s for which the probability that S exceeds s is less than $\frac{1}{2}$.

5. If the probability that a person lives past 90 years of age is $\frac{1}{50}$, what is the approximate probability that among a high school graduating class of 300 at most two live past 90 years of age? (Assume independent trials each with probability of success $\frac{1}{50}$. Comment on the assumption of independence.)

6. Let λ be a fixed positive number and consider $e^{-\lambda}(\lambda^k/k!)$ for nonnegative integer values of k. Find the value (or values) of k for which $e^{-\lambda}(\lambda^k/k!)$ is largest. [HINT: compute the ratio of $e^{-\lambda}(\lambda^r/r!)$ to $e^{-\lambda}(\lambda^{r+1}/(r+1)!)$, and determine for what choices of r this ratio is larger than one and for what choices of r this ratio is less than one.]

7. Let n be a fixed positive integer, and let p be a number between 0 and 1. Consider $\binom{n}{k} p^k(1-p)^{n-k}$ for $0 \le k \le n$ and find the value (or values) of k for which $\binom{n}{k} p^k(1-p)^{n-k}$ is largest. (Proceed as in the previous exercise.) Compare the answers to this problem and the previous one with $\lambda = np$.

*9.3 Random variables which assume a countably infinite number of values

The following discussion of random variables that may assume a countably infinite number of different values is, of course, outside the setting of the previous chapters—that of a finite sample space and a probability measure on it. We could enlarge that context to the extent that random variables with infinitely many values could be accommodated, as we did in Chapter Five, and then treat such topics as joint probability distributions, independence of random variables, a Chebychev inequality, etc., in analogous fashion to what we have done for random variables assuming finitely many values. Rather than do that here, however, we shall content ourselves with sketching the notions of probability distribution, expectation, and variance in the infinite case, illustrating the importance of these topics.

The parallelism we draw upon is quite direct in the case that the random variable in question assumes a countably infinite number of values. (Recall that a set S is countably infinite if its elements may be put into one-to-one correspondence with the positive integers 1, 2, 3, . . . , i.e., the elements of S can be enumerated in a list $s_1, s_2, s_3, \ldots$.) We shall therefore discuss a random variable X having possible values $x_1, x_2, x_3, \ldots$. Clearly, if X is to be a function on a sample space Ω, then Ω must itself be an infinite set. We shall skirt the problem of such a definition of X.

Let us first recall the following things from previous chapters. Suppose Ω is a *finite* sample space, P is a probability measure on Ω, and Y is a random variable on Ω. The random variable Y can assume only a finite number of distinct values, say $y_1, y_2, \ldots, y_m$, and, in particular, m is no larger than the number of sample points in Ω. The probability distribution f_Y of Y is given by $f_Y(y_i) = P(\{\omega | Y(\omega) = y_i\})$, $i = 1, 2, \ldots, m$. We then say the probability that $Y = y_i$ is $f_Y(y_i)$, and we have $f_Y(y_i) \geq 0$ and $\sum_{i=1}^{m} f_Y(y_i) = 1$. In terms of f_Y, we may compute, among other things, the probability that Y takes a value between two given numbers a and b, the expectation of Y, and the variance of Y.

We proceed formally by observing that we can specify the probability distribution of a random variable X, which is capable of assuming a countably infinite number of values, by specifying the probability $f_X(x_i)$ with which X takes the value x_i, $i = 1, 2, 3, \ldots$. Of course, since $f_X(x_i)$ is to be a probability, it should be nonnegative. Further, since it is certain that X assumes one of the values $x_1, x_2, x_3, \ldots$, we ask that the sum of the probabilities $f_X(x_i)$ be one. Thus, ignoring the question of a sample space on which to define X, we may nevertheless arrive at the formal conclusion that the probability distribution of X is specified by a list of values $x_1, x_2, x_3, \ldots$ together with the probabilities $f_X(x_1), f_X(x_2), f_X(x_3), \ldots$ of assuming them. These probabilities are subject to

$$f_X(x_i) \geq 0, \qquad i = 1, 2, 3, \ldots, \qquad \sum_{i=1}^{\infty} f_X(x_i) = 1 \tag{9.7}$$

The following examples of the above set-up might be viewed as theoretical. At the end of the section, we shall discuss some examples of the same type in more down-to-earth surroundings.

Example 9.3

The Geometric Distribution Suppose we have a sequence of Bernoulli trials with probability p of success in a given trial, $0 < p < 1$. Let X be the number of trials necessary to produce the *first* success. Alternatively, X is the waiting time for the first success.

Given *any* integer $k \geq 1$, it is conceivable that k trials are necessary in order to obtain the first success, i.e., there are $k - 1$ failures followed by a success. Thus, X may assume any of the values $1, 2, 3, \ldots$. Let us find the probability that X assumes the value k. In order that $X = k$, we must have $k - 1$ failures and then a success. Since the probability of failure in a given trial is $q = 1 - p$, and since the trials are independent, we see that the probability that $X = k$ is $q^{k-1}p$. Thus, the probability distribution of X can be described completely by saying X assumes the values $1, 2, 3, \ldots$ with respective probabilities $q^0p, q^1p, q^2p, \ldots$. In other words,

$$f_X(k) = q^{k-1}p, \qquad k = 1, 2, 3, \ldots \tag{9.8}$$

As defined in Eq. (9.8), f_X is indeed a probability distribution, i.e., f_X satisfies Eq. (9.7). Certainly $f_X(k) \geq 0$ for each k. As for the sum of the probabilities $f_X(k)$, we have

$$\sum_{k=1}^{\infty} f_X(k) = \sum_{k=1}^{\infty} q^{k-1}p = p \sum_{k=1}^{\infty} q^{k-1} \tag{9.9}$$

The series $\sum_{k=1}^{\infty} q^{k-1}$ is a geometric series whose sum is $(1 - q)^{-1}$. (Note that we have assumed that $0 < q < 1$, and this series converges.) Now $1 - q = p$, and so

$$\sum_{k=1}^{\infty} f_X(k) = p \cdot \frac{1}{p} = 1 \tag{9.10}$$

Since the terms pq^{k-1} are the successive terms of a geometric series, this distribution is often called the *geometric distribution*. A random variable having the probability distribution given by Eq. (9.8) is said to have a *geometric distribution with parameter* p.

Example 9.4

The Negative Binomial Distribution Let us continue with the same setting as in Example 9.3, but now letting r be a fixed positive integer and letting X be the number of trials necessary to produce the rth success. The preceding example is just the special case $r = 1$ of the present one.

Clearly, the rth success in Bernoulli trials cannot occur before the rth trial. This means that X cannot assume any value less than r. If we let k be a nonnegative integer, then X can assume the value $r + k$. Indeed, the event $X = r + k$ is precisely the event that in the first $r + k - 1$ trials there will be exactly $r - 1$ successes and then the $(r + k)$th trial is a success (which is the rth success). Since the trials are independent, the probability of this event is $\binom{r+k-1}{r-1} p^{r-1}q^k p$. That is,

$$f_X(r + k) = \binom{r+k-1}{r-1} p^r q^k, \qquad k = 0, 1, 2, \ldots \tag{9.11}$$

Clearly, $f_X(r + k) \geq 0$ for each k. To check that these probabilities add to one, we have

$$\begin{aligned} \sum_{k=0}^{\infty} f_X(r + k) &= \sum_{k=0}^{\infty} \binom{r+k-1}{r-1} p^r q^k \\ &= p^r \sum_{k=0}^{\infty} \binom{r+k-1}{r-1} q^k \end{aligned} \tag{9.12}$$

You can check that the series $\sum_{k=0}^{\infty} \binom{r+k-1}{r-1} q^k$ is precisely the binomial series for $(1 - q)^{-r}$. Thus, the sum is p^{-r}, and

$$\sum_{k=0}^{\infty} f_X(r + k) = p^r \cdot p^{-r} = 1 \tag{9.13}$$

Because the terms $\binom{r+k-1}{r-1} p^r q^k$ are the successive terms of the binomial expansion of $p^r(1 - q)^{-r}$, and because of the negative exponent $-r$, the probability distribution given in Eq. (9.11) is usually called the *negative binomial distribution.* A random variable with probability distribution (9.11) is said to have a *negative binomial distribution with parameters r and p.*

Example 9.5

The Poisson Distribution Suppose X is a random variable that assumes nonnegative integer values. Suppose further that for each large integer n, X is approximately the number of successes in n Bernoulli trials with probability λ/n of success in each trial. That is, suppose for each integer k

$$f_X(k) \cong \binom{n}{k}\left(\frac{\lambda}{n}\right)^k\left(1 - \frac{\lambda}{n}\right)^{n-k} \tag{9.14}$$

If the approximation claimed in Eq. (9.14) becomes increasingly good as n becomes large, we would be led, in view of Section 9.2, to the conclusion that

$$f_X(k) = e^{-\lambda}\frac{\lambda^k}{k!}, \qquad k = 0, 1, 2, \ldots \tag{9.15}$$

We have previously noted that f_X as defined by Eq. (9.15) is in fact a probability distribution, i.e., it satisfies Eq. (9.7).

We make no claim of rigor in the "derivation" of f_X above. The limiting approximation process that is described rather loosely is meant to be merely suggestive. A careful argument requires somewhat more delicacy, and we will be more explicit about it in a later example.

In the meantime, we remark that the distribution given in Eq. (9.15) is called the *Poisson distribution,* and a random variable having this distribution is said to be a *Poisson random variable with parameter* λ.

We next define the expectation of a random variable X that assumes a countably infinite number of values. If Y is a random variable that assumes a finite number of values, say y_i, with probability $f_Y(y_i)$, $i = 1, 2, \ldots, m$, then the expectation of Y is given by $\sum_{i=1}^{m} y_i f_Y(y_i)$. The analogous expression for the expectation of X would be $\sum_{i=1}^{\infty} x_i f_X(x_i)$. However, since this is now an infinite series, we cannot guarantee its convergence. We circumvent this difficulty with convergence by defining the expectation of X only in the case that $\sum_{i=1}^{\infty} x_i f_X(x_i)$ converges absolutely. Thus, if $\sum_{i=1}^{\infty} |x_i| f_X(x_i)$ converges, we define

$$E(X) = \sum_{i=1}^{\infty} x_i f_X(x_i) \tag{9.16}$$

Recall here that the absolute convergence of an infinite series guarantees the ordinary convergence of that series. Therefore, $\sum_{i=1}^{\infty} x_i f_X(x_i)$ is convergent. We shall denote this sum by μ.

For the most part, we are concerned with random variables that assume only nonnegative values. When X is such a random variable, there is no distinction between the ordinary and the absolute convergence of $\sum_{i=1}^{\infty} x_i f_X(x_i)$. The examples given below are of this kind and, consequently, we need then only test $\sum_{i=1}^{\infty} x_i f_X(x_i)$ for convergence.

When the expectation of X is defined—suppose it is μ—the variance of X is defined by

$$V(X) = \sum_{i=1}^{\infty} (x_i - \mu)^2 f_X(x_i) \tag{9.17}$$

Now, the series in Eq. (9.17) contains only nonnegative terms, so it either converges or it diverges to $+\infty$. When it is convergent, $V(X)$ may also be computed according to

$$\begin{aligned} V(X) &= \sum_{i=1}^{\infty} (x_i - \mu)^2 f_X(x_i) = \sum_{i=1}^{\infty} (x_i^2 - 2\mu x_i + \mu^2) f_X(x_i) \\ &= \sum_{i=1}^{\infty} x_i^2 f_X(x_i) - 2\mu \sum_{i=1}^{\infty} x_i f_X(x_i) + \mu^2 \sum_{i=1}^{\infty} f_X(x_i) \qquad (9.18) \\ &= \sum_{i=1}^{\infty} x_i^2 f_X(x_i) - \mu^2 \end{aligned}$$

In other words, $V(X)$ may be computed in the same way as was pointed out in Chapter Seven. When the series in Eq. (9.17) is divergent, X is simply said to have infinite variance.

If X has expectation μ and finite variance $V(X)$, these numbers admit an interpretation as location and dispersion parameters of the probability distribution f_X of X. We now compute them when X has successively a geometric, a negative binomial, and a Poisson distribution.

Example 9.6

Let X have a geometric distribution with parameter p, $0 < p < 1$, and, consider the expectation and the variance of X. We have, if the series converges,

$$E(X) = \sum_{k=1}^{\infty} kpq^{k-1} = p \sum_{k=1}^{\infty} kq^{k-1} = p(1 + 2q + 3q^2 + \cdots) \tag{9.19}$$

This series has the sum $(1 - q)^{-2} = p^{-2}$. [This may be seen by appealing to the binomial expansion with a negative exponent or, more directly, by observing that $(1 - q)(1 + 2q + 3q^2 + \cdots) = 1 + q + q^2 + \cdots = (1 - q)^{-1}$.] Thus,

$$E(X) = p \cdot p^{-2} = p^{-1} \tag{9.20}$$

Next,

$$\sum_{k=1}^{\infty} k^2 pq^{k-1} = p \sum_{k=1}^{\infty} [k(k+1) - k]q^{k-1}$$
$$= p \sum_{k=1}^{\infty} k(k+1)q^{k-1} - p \sum_{k=1}^{\infty} kq^{k-1} \tag{9.21}$$

The last series in Eq. (9.21) is just $E(X) = p^{-1}$, while

$$\sum_{k=1}^{\infty} k(k+1)q^{k-1} = (2 + 2\cdot 3q + 3\cdot 4q^2 + \cdots)$$

may be seen to be $2(1-q)^{-3}$ [observe that $(1-q)(2 + 2\cdot 3q + 3\cdot 4q^2 + \cdots) = 2 + 4q + 6q^2 + \cdots = 2(1-q)^{-2}$]. Therefore, in the light of Eq. (9.18),

$$V(X) = \sum_{k=1}^{\infty} k^2 pq^{k-1} - [E(X)]^2$$
$$= p \sum_{k=1}^{\infty} k(k+1)q^{k-1} - p \sum_{k=1}^{\infty} kq^{k-1} - p^{-2} \tag{9.22}$$
$$= 2p(1-q)^{-3} - p^{-1} - p^{-2} = p^{-2} - p^{-1} = qp^{-2}$$

For instance, the number of tosses of a balanced coin until a head appears has an expectation of $(\frac{1}{2})^{-1} = 2$ and a variance of $(\frac{1}{2})(\frac{1}{2})^{-2} = 2$.

Example 9.7

Let X have a negative binomial distribution with parameters r and p. Then

$$E(X) = \sum_{k=0}^{\infty} (r+k)\binom{r+k-1}{r-1} p^r q^k = p^r \sum_{k=0}^{\infty} r \binom{r+k}{r} q^k$$
$$= rp^r \sum_{k=0}^{\infty} \binom{r+k}{r} q^k = rp^r(1-q)^{-r-1} = rp^{-1} \tag{9.23}$$

where again we have identified $\sum_{k=0}^{\infty} \binom{r+k}{r} q^k$ as a binomial expansion of $(1-q)^{-r-1}$. Next,

$$\sum_{k=0}^{\infty} (r+k)^2 \binom{r+k-1}{r-1} p^r q^k$$
$$= \sum_{k=0}^{\infty} [(r+k)(r+k+1) - (r+k)]\binom{r+k-1}{r-1} p^r q^k \tag{9.24}$$
$$= p^r \sum_{k=0}^{\infty} r(r+1)\binom{r+k+1}{r+1} q^k - p^r \sum_{k=0}^{\infty} r \binom{r+k}{r} q^k$$

The last series in Eq. (9.24) is just $E(X) = rp^{-1}$, while

$$p^r \sum_{k=0}^{\infty} r(r+1)\binom{r+k+1}{r+1} q^k = r(r+1)p^r(1-q)^{-r-2}$$
$$= r(r+1)p^{-2}$$

Therefore,

$$\begin{aligned} V(X) &= \sum_{k=0}^{\infty} (r+k)^2 \binom{r+k-1}{r-1} p^r q^k - [E(X)]^2 \\ &= r(r+1)p^{-2} - rp^{-1} - r^2p^{-2} = r(p^{-2} - p^{-1}) = rqp^{-2} \end{aligned} \tag{9.25}$$

Here we have found the expectation and variance of the negative binomial distribution, and the results, with $r = 1$, agree with Eqs. (9.20) and (9.22).

Example 9.8

Finally, let X have a Poisson distribution with parameter λ. Then,

$$E(X) = \sum_{k=0}^{\infty} k \frac{\lambda^k}{k!} e^{-\lambda} = e^{-\lambda} \sum_{k=0}^{\infty} k \frac{\lambda^k}{k!} \tag{9.26}$$

The term corresponding to $k = 0$ of the series in Eq. (9.26) is zero, so the sum can be extended beyond $k = 1$. Hence,

$$\begin{aligned} E(X) &= e^{-\lambda} \sum_{k=1}^{\infty} k \frac{\lambda^k}{k!} = e^{-\lambda} \sum_{k=1}^{\infty} \frac{\lambda^k}{(k-1)!} \\ &= \lambda e^{-\lambda} \sum_{k=1}^{\infty} \frac{\lambda^{k-1}}{(k-1)!} = \lambda e^{-\lambda} e^{\lambda} = \lambda \end{aligned} \tag{9.27}$$

Thus, the parameter λ is the expectation of X. (We will refer to this fact later.)

Now let us compute $V(X)$. We have

$$\begin{aligned} \sum_{k=0}^{\infty} k^2 e^{-\lambda} \frac{\lambda^k}{k!} &= e^{-\lambda} \sum_{k=0}^{\infty} [k(k-1) + k] \frac{\lambda^k}{k!} \\ &= e^{-\lambda} \sum_{k=0}^{\infty} k(k-1) \frac{\lambda^k}{k!} + e^{-\lambda} \sum_{k=0}^{\infty} k \frac{\lambda^k}{k!} \end{aligned} \tag{9.28}$$

The last series in Eq. (9.28) is $E(X) = \lambda$, while $\sum_{k=0}^{\infty} k(k-1)(\lambda^k/k!)$ may be seen to be $\sum_{k=2}^{\infty} k(k-1)(\lambda^k/k!)$. Hence, we have

$$\begin{aligned} V(X) &= \sum_{k=0}^{\infty} k^2 e^{-\lambda} \frac{\lambda^k}{k!} - \lambda^2 = e^{-\lambda} \sum_{k=2}^{\infty} k(k-1) \frac{\lambda^k}{k!} + \lambda - \lambda^2 \\ &= e^{-\lambda} \sum_{k=2}^{\infty} \frac{\lambda^k}{(k-2)!} + \lambda - \lambda^2 \\ &= \lambda^2 e^{-\lambda} \sum_{k=2}^{\infty} \frac{\lambda^{k-2}}{(k-2)!} + \lambda - \lambda^2 \\ &= \lambda^2 e^{-\lambda} e^{\lambda} + \lambda - \lambda^2 = \lambda \end{aligned} \tag{9.29}$$

A random variable having a Poisson distribution with parameter λ is then seen to have both an expectation and a variance of λ.

In conclusion, we consider some applied settings to which the above discussion is clearly relevant. To begin with, there are situations in which the geometric or, more generally, the negative binomial distribution arises in a natural way. Indeed, it is not difficult to imagine performing a sequence of Bernoulli trials until the first, or the rth, success is obtained. To indicate for what purpose this might be done, let us consider the following important application to the problem of estimating the size of a wildlife population.

Example 9.9

Suppose there are N deer in a preserve (N is unknown), of which t have been affixed with tags of some sort (t is known). Now consider the experiment of capturing a random sample of size one from the population. Let a success be scored if the captured animal is tagged. Independent replications of this experiment performed with replacement until the rth success is obtained would require X captures. The random variable X would then have a negative binomial distribution with parameters r and t/N (unknown). In this, we are clearly avoiding the very real problem of capturing such a random sample. For example, there may be deer that are especially prone to capture.

Within the mathematical model, the magnitude of X will provide an indication of the size of N. On the one hand, if N is large in relation to t so that the probability of success in a given trial t/N is small, X is likely to be large also. On the other hand, if N is not much larger than t so that t/N is not much smaller than one, then X is likely to be small.

To be more precise, we have found the expectation of X to be rN/t and the variance of X to be $(rN^2/t^2)(1 - t/N)$. You may check that the random variable tX/r has expectation N and variance $(N^2/r)(1 - t/N)$. It is common in this circumstance to adopt tX/r as an estimate of the population size N. Further important properties of this estimate derive from a Chebychev inequality (see Exercise 12) and a central limit theorem. The details of this are omitted.

There are also many applied problems in which the random variable of interest might reasonably be supposed to be a Poisson random variable. The first type of problem of this sort is one in which the random variable in question can be thought of as being the number of successes in n Bernoulli trials with probability of success p in each trial, n large and p small. It is not unreasonable then to adopt the Poisson distribution with $\lambda = np$ as an approximate model. For example, consider a rare noncommunicable disease for which the probability that a person in the United States succumbs to an attack within a year's time is 1/50,000,000. Then the number of fatal cases within a year's time might be viewed as the result of, say, 200,000,000 Bernoulli trials each with probability of "success" 1/50,000,000. We could then expect that the Poisson distribution with parameter $\lambda = 4$ would serve as a very adequate model. That is, the probability of having k fatal cases within a year is, to a good approximation, $e^{-4}(4^k/k!)$. Notice that the Poisson distribution assigns positive probability to having 400,000,000 fatal cases; however, this is an exceedingly small probability, and, therefore, it is close to the correct probability of zero.

Another possible reason to look toward the Poisson distribution as a probability distribution for some specific random-variable is expanded upon in the following example.

Example 9.10

A radioactive substance decays by emitting α-particles, and the number of particles emitted during a total of s seconds can be counted by means of sophisticated Geiger counters. There is an inherent randomness in this process, and we would like to consider the random variable X, which represents the number of emissions. Clearly, X takes nonnegative integer values. Here is a "justification" for taking X to be a Poisson random variable.

Imagine the s seconds divided up into a large number, say n, of segments of s/n seconds each. In any one short segment, there is either no emission or there is one or more emissions. If n is very large so that s/n is very small, the possibility of having two or more emissions might be ignored as being unlikely relative to having no emissions or one emission in the small interval. If we do this, X is approximately the number of time segments of length s/n in which there is one emission. Now there is some basis for assuming that emissions in different intervals are independent events. With these assumptions, X is approximately the number of successes (a success means at least one emission) in n Bernoulli trials (time segments). We are here close to the set-up of Example 9.5.

Let us suppose, as a final assumption, that the probability of having at least one emission during a small time interval is approximately a fixed proportion of the length of that interval. That is, let us suppose that for n large and s/n small, the probability of a success in an interval of s/n seconds is approximately λ/n. We then reach the conclusion that the probability that $X = k$ is about $e^{-\lambda}(\lambda^k/k!)$. If our assumptions are reasonable, then X can be taken to be a Poisson random variable with parameter λ.

The significance of the parameter λ is that it represents the expected number of particles emitted in s seconds time. Thus, λ is characteristic of the rate at which the substance decays. The faster is this rate, the higher is λ. Of course, λ will be proportional to the length of the time interval, s. If the observation were based on a time interval of different length, we would get a different value for λ. We have, in fact, supposed that the rate of emission is constant throughout the s second interval. This may not be the case, and the emission rate may be, for example, slowing down. However, many substances decay so slowly that the rate might well be taken as constant for a period of hours or days.

For a given substance, we might estimate the value of λ as follows. Suppose we observe the substance over a large number, say N, of time intervals of s seconds each and record how many particles were emitted in each of these N intervals. Then the sample average count would provide us with an estimate of λ, the expectation of the number of particles emitted during s seconds. If N_k is the number of time intervals observed during which exactly k particles were emitted, then this average can be computed to be $\sum kN_k/N$. Further, N_k/N is the relative frequency of occurrence of the event that exactly k particles are emitted.

If the model described above is a good one, we would expect that if such a substance were observed during N time intervals of s seconds each, and if λ were estimated by computing the average number of particles emitted, then the relative frequencies N_k/N should be close to $e^{-\lambda}(\lambda^k/k!)$, the probability of exactly k particles being emitted during a given time interval according to the model described above.

In a famous experiment performed in 1919, Rutherford, Chadwick, and Ellis observed a radioactive substance over $N = 2608$ time intervals of $s = 7.5$ seconds each and recorded the number N_k of time intervals during which exactly k particles were emitted. The values they observed are given in Table 9.1. The total number of particles emitted during the

TABLE 9.1

Radioactive emission, $N = 2608$, $\lambda = 3.87$

k	N_k	$Ne^{-\lambda}\dfrac{\lambda^k}{k!}$
0	57	54.399
1	203	210.523
2	383	407.361
3	525	525.496
4	532	508.418
5	408	393.515
6	273	253.817
7	139	140.325
8	45	67.882
9	27	29.189
$k \geq 10$	10	17.075
TOTAL	$N = 2608$	2608.000

N time intervals was $\sum kN_k = 10{,}094$, and the average number was therefore $10{,}094/2608 = 3.87$. Thus, an estimate for λ is 3.87. If the model of the Poisson distribution is a good one for this process of radioactive emission, we would expect N_k/N to be close to $e^{-\lambda}(\lambda^k/k!)$, or, in other words, N_k to be close to $Ne^{-\lambda}(\lambda^k/k!)$. From $\lambda = 3.87$, the values of $Ne^{-\lambda}(\lambda^k/k!)$ have been computed for each k and compared with the observed values N_k. (The results are in Table 9.1.) It may be observed that the observed values N_k are rather close to the theoretical values $Ne^{-\lambda}(\lambda^k/k!)$.

The basic point in Example 9.10 is the division of time into small segments within which we supposed independent events. There are numerous other examples of the same type in which the Poisson distribution provides a sound model. In particular, the same assumptions we have invoked are also applied to counts of some sort in a given area or volume as opposed to a time segment. If you wish to examine more problems of this type, refer to the book of Feller, cited at the end of this book.

EXERCISES 9.3

1. Two balanced dice are thrown. In a sequence of Bernoulli trials of this experiment, what is the probability distribution of the number of trials necessary

to produce the first total score of six? What is the expectation of this number of trials?

2. In Exercise 1, what is the probability distribution of the number of trials required to find at least one die with a score of three? What is the expectation and variance of this number of trials?

3. In Exercise 1, what is the probability distribution of the number of trials required to produce the first total score of six or seven? What is the probability that the first such total score is six?

4. In Exercise 1, what is the expectation and the variance of the number of trials until the scores on the two dice agree for the tenth time?

5. Suppose X has a geometric distribution with parameter p. What is the probability that X is no larger than a positive integer n? What is the probability that X is even and no larger than $2n$?

6. Suppose X is a random variable that assumes the value $2^{k/2}$ with probability $1/2^k$, $k = 1, 2, \ldots$. Find the expectation of X, and show that the variance of X is infinite.

7. Suppose X is a random variable that assumes the value 2^k with probability $1/2^{k+1}$ and the value -2^k with probability $1/2^{k+1}$, $k = 1, 2, \ldots$. Show that this is a probability distribution. Is the expectation of X defined?

8. Write an expression for the probability that a Poisson random variable with parameter λ assumes an even value no larger than $2n$ for a positive integer n. What is the limit of this expression as $n \to \infty$?

9. Let X be a random variable that assumes the value x_i with probability $f_X(x_i)$, $i = 1, 2, \ldots$, and let g be a function of a real variable. By $g(X)$ we mean the random variable that assumes the value $g(x_i)$ when $X = x_i$. Show that

   (a) $E(cX) = cE(X)$ when $\sum_{i=1}^{\infty} |x_i| f_X(x_i)$ converges.

   (b) $E(X + c) = E(X) + c$ when $\sum_{i=1}^{\infty} |x_i| f_X(x_i)$ converges.

   (c) $V(cX) = c^2 V(X)$ when X has a finite variance.
   (d) $V(X + c) = V(X)$ when X has a finite variance.

*10. Show generally that $E(g(X)) = \sum_{i=1}^{\infty} g(x_i) f_X(x_i)$ provided $\sum_{i=1}^{\infty} |g(x_i)| f_X(x_i)$ converges.

11. Let X be a Poisson random variable with parameter λ. Compute $E(X(X - 1) \cdots (X - n + 1))$.

*12. Suppose X has a negative binomial distribution with parameters r and p. Show by duplicating the proof of the Chebychev inequality for the case of finitely many values that a corresponding inequality holds for X. Interpret this result in the context of Example 9.9.

APPENDIX

The Normal Distribution Function

The function defined by

$$\varphi(x) = \frac{e^{-x^2/2}}{\sqrt{2\pi}} \tag{A.1}$$

for real numbers x, is called the *normal density function*. The function defined by

$$\Phi(t) = \int_{-\infty}^{t} \frac{e^{-x^2/2}}{\sqrt{2\pi}}\, dx = \frac{1}{\sqrt{2\pi}} \int_{-\infty}^{t} e^{-x^2/2}\, dx \tag{A.2}$$

for real numbers t, is called the *normal distribution function*. In this appendix we collect some useful facts about the functions φ and Φ.

Let us begin with φ. First, we note that $e^{-x^2/2}$ is positive for any real number x and therefore that φ is a positive function. Second, we observe that the function φ possesses a symmetry property that arises from the fact that $\varphi(x) = \varphi(-x)$ for any real number x. Third, we see that $\varphi(x) \to 0$ as $x \to \infty$. These properties of φ, as well as a general "bell-shape," are manifested in the graph of φ, which is plotted in Figure A1.

Now we consider the normal distribution function Φ. $\Phi(t)$ is just the definite integral of φ between the limits $-\infty$ and t, and, as such, admits an interpretation as the area under the graph of φ to the left of t. See Figure A2.

Inasmuch as φ is a positive function, $\Phi(t) > 0$ for each real t, and, as t is increased, $\Phi(t)$ also increases. When $t \to -\infty$, $\Phi(t) \to 0$, which is usually paraphrased by writing $\Phi(-\infty) = 0$. It is an important fact about the function Φ, that $\Phi(t) \to 1$ as $t \to +\infty$. This fact is expressed by writing $\Phi(+\infty) = 1$. In terms of area, $\Phi(+\infty) = 1$ means that the total area under the graph of φ is equal to 1. Thus, Φ is a positive and increasing function of t which increases from 0 to 1 as t increases from $-\infty$ to $+\infty$. The graph of Φ is plotted in Figure A3.

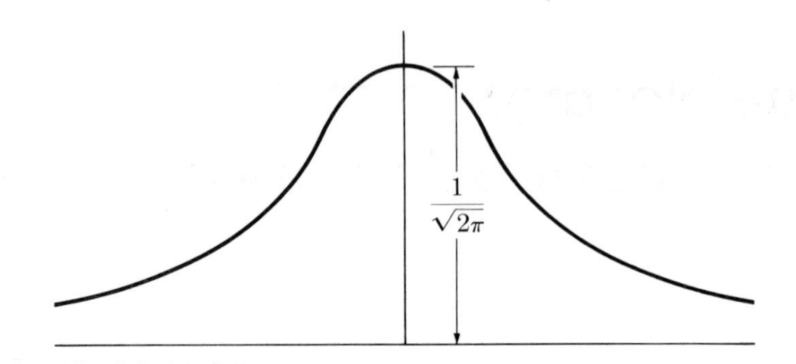

FIGURE A.1

Graph of $\varphi(x) = \dfrac{1}{\sqrt{2\pi}} e^{-x^2/2}$

To see that $\Phi(+\infty) = 1$, let us set $K = \Phi(+\infty)$. Then

$$\begin{aligned} K^2 &= \left(\frac{1}{\sqrt{2\pi}} \int_{-\infty}^{+\infty} e^{-x^2/2}\, dx\right)\left(\frac{1}{\sqrt{2\pi}} \int_{-\infty}^{+\infty} e^{-y^2/2}\, dy\right) \\ &= \frac{1}{2\pi} \int_{-\infty}^{+\infty} \int_{-\infty}^{+\infty} e^{-x^2/2} e^{-y^2/2}\, dx\, dy \qquad \text{(A.3)} \\ &= \frac{1}{2\pi} \int_{-\infty}^{+\infty} \int_{-\infty}^{+\infty} e^{-(x^2+y^2)/2}\, dx\, dy \end{aligned}$$

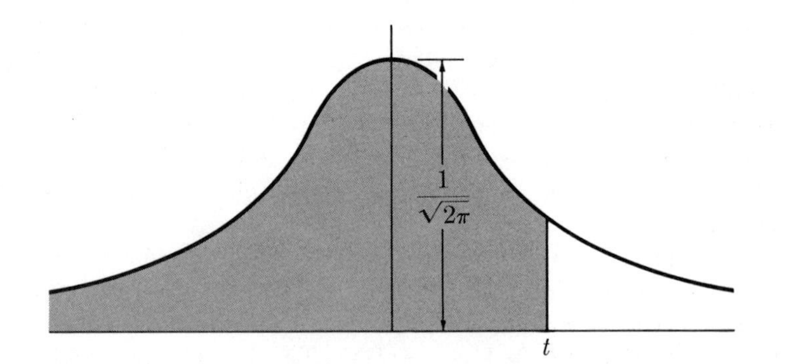

FIGURE A.2

Shaded area equals $\Phi(t) = \dfrac{1}{\sqrt{2\pi}} \displaystyle\int_{-\infty}^{t} e^{-x^2/2}\, dx$

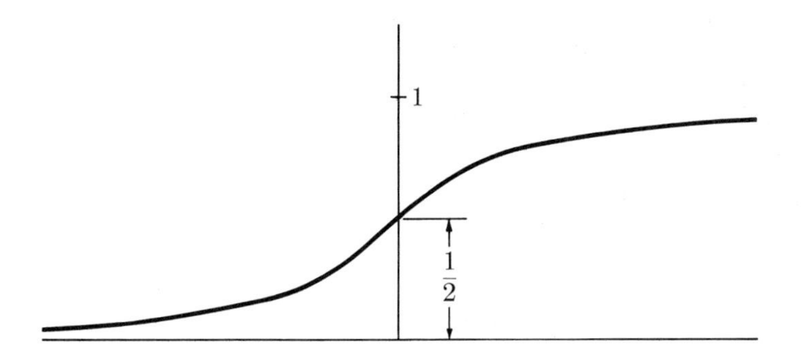

FIGURE A.3
Graph of $\Phi(t) = \dfrac{1}{\sqrt{2\pi}} \displaystyle\int_{-\infty}^{t} e^{-x^2/2}\, dx$

We can regard the double integral in Eq. (A.3) as an integral of the function $e^{-(x^2+y^2)/2}$ over the (x, y)-plane. If we now transform to polar coordinates (r, θ), we have $x^2 + y^2 = r^2$ and $dx\, dy = r\, dr\, d\theta$. The variable r ranges from 0 to $+\infty$ and θ from 0 to 2π. Thus,

$$K^2 = \frac{1}{2\pi} \int_0^{2\pi} \int_0^{+\infty} e^{-r^2/2} r\, dr\, d\theta \tag{A.4}$$

The substitution $r^2/2 = z$ reduces $\int_0^{+\infty} e^{-r^2/2} r\, dr$ to $\int_0^{+\infty} e^{-z}\, dz$, which is equal to 1. Thus,

$$K^2 = \frac{1}{2\pi} \int_0^{2\pi} 1\, d\theta = 1 \tag{A.5}$$

From Eq. (A.5), we see that K, which is obviously positive, must equal 1 as must $\Phi(+\infty)$.

We now turn to the question of evaluating $\Phi(t)$ for various values of t. Since $\Phi(t)$ is given by Eq. (A.2), our problem would be simple if we could perform the indicated integration and express the result in closed form in terms of elementary functions. Unfortunately, this cannot be done. The best we can do is to resort to numerical approximations to the values of $\Phi(t)$ for different values of t. Tables are available for this purpose, and one such table is reproduced in Table A.1.

Table A.1 gives approximate values of $\Phi(t)$ for a range of nonnegative values of t. Thus, if we enter the row of the table labelled 1.3 and go to the column labelled .06, we find directly that $\Phi(1.36) \sim .9131$. Again, entering the row labelled .7 and going to the column marked .02, we find $\Phi(.72) \sim .7643$. For t larger than 3.49, we have $\Phi(t) \sim 1$ inasmuch as

$\Phi(t)$ must exceed $\Phi(3.49)$ which is about .9998. If, on the other hand, t takes a value between two tabulated values, say $t = 1.034$, we notice that $.8485 \sim \Phi(1.03) < \Phi(1.034) < \Phi(1.04) \sim .8508$.

Table A.1 may also be used to find an approximate value of $\Phi(t)$ when $t < 0$. In fact, using the symmetry property of φ, $\varphi(x) = \varphi(-x)$ for any real x, it follows that

$$\Phi(t) = \int_{-\infty}^{t} \varphi(x)\, dx = \int_{-t}^{\infty} \varphi(x)\, dx = 1 - \Phi(-t) \qquad \text{(A.6)}$$

In this regard, see Figure A4. Now in order to find approximately $\Phi(-2.35)$, for example, we first determine that $\Phi(2.35) \sim .9906$ and so $\Phi(-2.35) = 1 - \Phi(2.35) \sim .0094$.

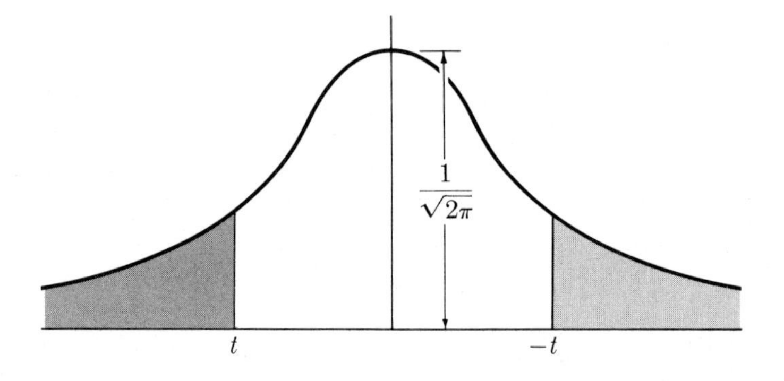

FIGURE A.4
Dark shaded area equals $\Phi(t)$, lighter shaded area equals $1 - \Phi(-t)$, $t < 0$

TABLE A.1

Values of $\Phi(t) = \frac{1}{\sqrt{2\pi}} \int_{-\infty}^{t} e^{-x^2/2}\, dx$ for nonnegative t

	.00	.01	.02	.03	.04	.05	.06	.07	.08	.09
.0	.5000	.5040	.5080	.5120	.5160	.5199	.5239	.5279	.5319	.5359
.1	.5398	.5438	.5478	.5517	.5557	.5596	.5636	.5675	.5714	.5753
.2	.5793	.5832	.5871	.5910	.5948	.5987	.6026	.6064	.6103	.6141
.3	.6179	.6217	.6255	.6293	.6331	.6368	.6406	.6443	.6480	.6517
.4	.6554	.6591	.6628	.6664	.6700	.6736	.6772	.6808	.6844	.6879
.5	.6915	.6950	.6985	.7019	.7054	.7088	.7123	.7157	.7190	.7224
.6	.7257	.7291	.7324	.7357	.7389	.7422	.7454	.7486	.7517	.7549
.7	.7580	.7611	.7643	.7673	.7704	.7734	.7764	.7794	.7823	.7852
.8	.7881	.7910	.7939	.7967	.7995	.8023	.8051	.8078	.8106	.8133
.9	.8159	.8186	.8212	.8238	.8264	.8289	.8315	.8340	.8365	.8389
1.0	.8413	.8438	.8461	.8485	.8508	.8531	.8554	.8577	.8599	.8621
1.1	.8643	.8665	.8686	.8708	.8729	.8749	.8770	.8790	.8810	.8830
1.2	.8849	.8869	.8888	.8907	.8925	.8944	.8962	.8980	.8997	.9015
1.3	.9032	.9049	.9066	.9082	.9099	.9115	.9131	.9147	.9162	.9177
1.4	.9192	.9207	.9222	.9236	.9251	.9265	.9279	.9292	.9306	.9319
1.5	.9332	.9345	.9357	.9370	.9382	.9394	.9406	.9418	.9429	.9441
1.6	.9452	.9463	.9474	.9484	.9495	.9505	.9515	.9525	.9535	.9545
1.7	.9554	.9564	.9573	.9582	.9591	.9599	.9608	.9616	.9625	.9633
1.8	.9641	.9649	.9656	.9664	.9671	.9678	.9686	.9693	.9699	.9706
1.9	.9713	.9719	.9726	.9732	.9738	.9744	.9750	.9756	.9761	.9767
2.0	.9772	.9778	.9783	.9788	.9793	.9798	.9803	.9808	.9812	.9817
2.1	.9821	.9826	.9830	.9834	.9838	.9842	.9846	.9850	.9854	.9857
2.2	.9861	.9864	.9868	.9871	.9875	.9878	.9881	.9884	.9887	.9890
2.3	.9893	.9896	.9898	.9901	.9904	.9906	.9909	.9911	.9913	.9916
2.4	.9918	.9920	.9922	.9925	.9927	.9929	.9931	.9932	.9934	.9936
2.5	.9938	.9940	.9941	.9943	.9945	.9946	.9948	.9949	.9951	.9952
2.6	.9953	.9955	.9956	.9957	.9959	.9960	.9961	.9962	.9963	.9964
2.7	.9965	.9966	.9967	.9968	.9969	.9970	.9971	.9972	.9973	.9974
2.8	.9974	.9975	.9976	.9977	.9977	.9978	.9979	.9979	.9980	.9981
2.9	.9981	.9982	.9982	.9983	.9984	.9984	.9985	.9985	.9986	.9986
3.0	.9987	.9987	.9987	.9988	.9988	.9989	.9989	.9989	.9990	.9990
3.1	.9990	.9991	.9991	.9991	.9992	.9992	.9992	.9992	.9993	.9993
3.2	.9993	.9993	.9994	.9994	.9994	.9994	.9994	.9995	.9995	.9995
3.3	.9995	.9995	.9995	.9996	.9996	.9996	.9996	.9996	.9996	.9997
3.4	.9997	.9997	.9997	.9997	.9997	.9997	.9997	.9997	.9997	.9998

Suggestions for Further Reading

The student who wishes to pursue the study of probability further than this textbook will find several books available for additional reading. Among them, a treatment of discrete probability that is at a level somewhat more advanced than *Discrete Probability*, is

FELLER, WILLIAM. *An Introduction to Probability Theory and Its Applications*, 2nd ed., Vol. 1. New York: Wiley, 1957.

Feller's entertaining treatment of probability theory presumes that the reader has facility with the calculus and requires a good deal of mathematical sophistication.

There are also available several books at a mathematical level intermediate between Feller's and *Discrete Probability*. Among them are the following two books, both of which require a familiarity with the calculus and which cover both discrete and continuous probability.

CRAMÉR, HARALD. *The Elements of Probability Theory and Some of Its Applications*. New York: Wiley, 1955.

PARZEN, EMANUEL. *Modern Probability Theory and Its Applications*. New York: Wiley, 1960.

Books oriented toward statistical applications of probability theory are quite numerous. The following three will provide the interested student with a treatment at a relatively elementary level.

CHERNOFF, HERMAN, and LINCOLN E. MOSES. *Elementary Decision Theory*. New York: Wiley, 1959.

HODGES, J. L., JR., and E. L. LEHMANN. *Basic Concepts of Probability and Statistics*. San Francisco: Holden-Day, 1964.

NEYMAN, J. *First Course in Probability and Statistics*. New York: Holt, 1950.

Answers to Selected Exercises

Section 1.3

6. (**a**), (**c**), (**d**), (**f**), and (**g**) are true.
8. (**a**) $A = \varnothing$, (**b**) $B = \Omega$, (**c**) $A = \varnothing$ and $B = \Omega$.

Section 1.4

4. (**a**), (**b**), (**d**), and (**f**) are true.
11. Show that the complements of the two sets are equal.
12. (**a**) $N(A) = 0$, (**b**) $N(A \cap B) = 0$, (**c**) $0 < N(A) < N(\Omega)$, (**d**) $N(A \cap B') = 0$, (**e**) $N(A \cap B') + N(A' \cap B) = 0$.
13. $N(\Omega) = 49$, $N(A) = 26$, $N(A \cap B) = 14$, $N(A \cup B) = 34$, $N(A') = 23$, $N(B') = 27$, and $N(A' \cup B') = 35$.

Section 1.5

4. (**a**), (**b**), and (**c**) are true. **8.** $2^{m+n} - 2^m - 2^n$.

Section 2.5

2. (**a**) $A \cap B' \cap C'$, (**b**) $(A \cap B \cap C)'$, (**c**) $A \cap B \cap C$, (**e**) $A' \cap B' \cap C'$, (**f**) $(A \cap B \cap C') \cup (A \cap B' \cap C) \cup (A' \cap B \cap C)$, (**g**) $(A \cap B' \cap C') \cup (A' \cap B \cap C')$, (**h**) $A' \cap B$, (**i**) $B' \cap C'$, (**j**) $A \cup (B' \cap C')$.
5. (**a**) $A_2 \cap A_4'$, (**b**) $A_2 \cup A_5'$, (**c**) $A_3 \cap A_5$, (**d**) $A_2 \cap A_3$.

Section 3.2

3.

	$P(A')$	$P(A' \cap B)$	$P(A \cap B')$	$P(A \cup B)$	$P(A' \cap B')$
(**a**)	$\frac{2}{3}$	$\frac{1}{12}$	$\frac{1}{6}$	$\frac{5}{12}$	$\frac{7}{12}$
(**b**)	$\frac{1}{2}$	0	$\frac{3}{8}$	$\frac{1}{2}$	$\frac{1}{2}$
(**c**)	$\frac{1}{2}$	$\frac{1}{8}$	$\frac{1}{2}$	$\frac{5}{8}$	$\frac{3}{8}$
(**d**)	0	0	$\frac{3}{4}$	1	0

5. **(a)** $P(A \cup B) \geq \frac{2}{3}$, **(b)** $\frac{1}{6} \leq P(A \cap B) \leq \frac{1}{2}$, **(c)** $0 \leq P(A \cap B') \leq \frac{1}{3}$.

6. Apply Proposition 3.4 to A and $B \cup C$ to get $P(A \cup B \cup C) = P(A) + P(B \cup C) - P(A \cap (B \cup C))$. By another application of the same proposition, $P(B \cup C) = P(B) + P(C) - P(B \cap C)$. Finally, note that $P(A \cap (B \cup C)) = P((A \cap B) \cup (A \cap C)) = P(A \cap B) + P(A \cap C) - P(A \cap B \cap A \cap C) = P(A \cap B) + P(A \cap C) - P(A \cap B \cap C)$, by yet another application of Proposition 3.4.

8. Use induction and the fact that $P(A \cup B) \leq P(A) + P(B)$ for any two events A and B.

12. For **(a)**, let D be the union of all events that are P-null, and, using Exercise 8, show that D is P-null. For **(b)** and **(c)**, show that D' is contained in every P-sure event.

Section 3.4

2. **(a)** $\frac{1}{2}$, **(b)** $\frac{1}{4}$, **(c)** $\frac{1}{18}$, **(d)** $\frac{5}{12}$, **(e)** $\frac{1}{6}$.

4. **(a)** $\frac{1}{4}$, **(b)** $\frac{1}{20}$, **(c)** $\frac{1}{5}$, **(d)** $\frac{2}{25}$, **(e)** $\frac{3}{4}$, **(f)** $\frac{3}{5}$, **(g)** $\frac{2}{5}$.

6. 10, 6.

8. At least 5.

Section 3.5

2. **(a)** $\frac{25}{102}$, **(b)** $\frac{13}{51}$, **(c)** $\frac{77}{102}$, **(d)** $\frac{8}{663}$, **(e)** $\frac{1}{34}$.

4. **(a)** $4\,\dfrac{\binom{13}{5}}{\binom{52}{5}}$, **(b)** $\dfrac{\binom{4}{2}\binom{4}{3}}{\binom{52}{5}}$, **(c)** $13 \cdot 12\,\dfrac{\binom{4}{2}\binom{4}{3}}{\binom{52}{5}}$, **(d)** $4^5\,\dfrac{\binom{13}{5}}{\binom{52}{5}}$.

6. **(a)** $\binom{5}{3}\left(\frac{1}{6}\right)^3\left(\frac{5}{6}\right)^2$, **(b)** $\binom{5}{3}\left(\frac{1}{6}\right)^5$, **(c)** $\left(\frac{1}{6}\right)^5 + \binom{5}{1}\left(\frac{1}{6}\right)^5$, **(d)** $6!/6^5$.

8. $\dfrac{\binom{g}{2}}{\binom{s}{2}}, \dfrac{\binom{g}{2} + \binom{r}{2}}{\binom{s}{2}}$. **10.** $\dfrac{\binom{r-1}{w-1}}{\binom{t}{w}}, \dfrac{\binom{r-s-1}{w-2}}{\binom{t}{w}}$.

12. **(a)** $\dfrac{\binom{80}{10}}{\binom{200}{10}}$, **(b)** $\dfrac{\binom{140}{10}}{\binom{200}{10}}$, **(c)** $\dfrac{\binom{80}{4}\binom{60}{3}\binom{40}{2}\binom{20}{1}}{\binom{200}{10}}$, **(d)** $\dfrac{\binom{180}{10}}{\binom{200}{10}}$.

16. **(a)** $4\,\dfrac{\binom{48}{13,\,13,\,13}}{\binom{52}{13,\,13,\,13}}$, **(b)** $\dfrac{4!}{\binom{52}{13,\,13,\,13}}$, **(c)** $\dfrac{\binom{4}{1,\,1,\,1}\binom{48}{12,\,12,\,12}}{\binom{52}{13,\,13,\,13}}$.

18. $(n-1)!$, $2/(n-1)$. **20.** $\binom{n}{k}\left(\frac{1}{2}\right)^n$.

22. **(a)** $\binom{8}{2}\left(\frac{1}{26}\right)^2\left(\frac{25}{26}\right)^6$, **(b)** $1 - \left(\frac{25}{26}\right)^8$, **(c)** $\binom{8}{1}\left(\frac{1}{26}\right)\left(\frac{25}{26}\right)^7 - \binom{8}{1}\left(\frac{1}{26}\right)\left(\frac{24}{26}\right)^7$, **(d)** $1 - 2\left(\frac{25}{26}\right)^8 + \left(\frac{24}{26}\right)^8$, **(e)** $\left(\frac{1}{26}\right)^3$.

24. $\binom{100}{20}\left(\frac{1}{6}\right)^{20}\left(\frac{5}{6}\right)^{80}$.

Section 4.2

2. **(a)** $\frac{8}{11}$, **(b)** $\frac{8}{13}$, **(c)** $\frac{5}{9}$, **(d)** 1, **(e)** $\frac{11}{14}$.

4. $P_A(B) = 1$, $P_A(C) = 1 - \dfrac{\binom{48}{12}}{\binom{51}{12}}$, $P_B(A) = \dfrac{\binom{51}{12}}{\binom{52}{13} - \binom{48}{13}}$,

$P_B(C) = 1 - \dfrac{\binom{4}{1}\binom{48}{12}}{\binom{52}{13} - \binom{48}{13}}$, $P_C(A) = \dfrac{\binom{51}{12} - \binom{48}{12}}{\binom{52}{13} - \binom{48}{13} - \binom{4}{1}\binom{48}{12}}$,

$P_C(B) = 1$.

6. **(a)** $\frac{1}{6}$, **(b)** $\frac{1}{2}$, **(c)** $\frac{1}{6}$, **(d)** $\frac{1}{6}$, **(e)** 0, **(f)** $\frac{3}{5}$, **(g)** $\frac{3}{4}$, **(h)** $\frac{6}{7}$.

8. $(n/(n-1))p_n$, p_n as given in Example 3.10.

16. $\dbinom{k-k_1}{k_2}(1/(q-1))^{k_2}(1-(1/(q-1)))^{k-k_1-k_2}$,

$\dbinom{k-k_1}{k_2, k_3, \ldots, k_{q-1}}(1/(q-1))^{k-k_1}$.

Section 4.3

2.

	$P_{A'}(B_1)$	$P_{A'}(B_2)$	$P_{A'}(B_3)$	$P_{A'}(B_4)$
(a)	$\frac{1}{2}$	$\frac{1}{3}$	$\frac{1}{6}$	0
(b)	$\frac{3}{8}$	$\frac{1}{4}$	$\frac{3}{8}$	0
(c)	$\frac{3}{5}$	$\frac{3}{10}$	$\frac{1}{10}$	0

4. $\frac{3}{4}$.

6. $\frac{1500}{22}\%$, 22%.

Section 4.4

4. $P(A \cup B) = \frac{3}{8}$, $P(A' \cap B) = \frac{5}{24}$, $P(A' \cup B') = \frac{23}{24}$.

6. Independent events in **(c)**, **(d)**, and **(e)**.

8. Independent events in **(a)**, **(b)**, and **(e)**.

16. To show the statement of Exercise 15 is not true with pairwise independence replacing mutual independence, recall that the events A, B, and C of Exercise 11 are pairwise independent, then check that $A \cap B$ and C are dependent events.

Section 5.4

2.

X \ W	0	1	2	3	4
0	0	0	0	0	$\frac{1}{16}$
1	0	0	0	$\frac{1}{16}$	$\frac{3}{16}$
2	0	0	$\frac{1}{16}$	$\frac{2}{16}$	$\frac{3}{16}$
3	0	$\frac{1}{16}$	$\frac{1}{16}$	$\frac{1}{16}$	$\frac{1}{16}$
4	$\frac{1}{16}$	0	0	0	0

4. $f_{X_1}(r) = \binom{3}{r}\left(\frac{1}{3}\right)^r\left(\frac{2}{3}\right)^{3-r}, 0 \le r \le 3;$

$f_{X_2}(r) = \binom{3}{r}\left(\frac{1}{3}\right)^r\left(\frac{2}{3}\right)^{3-r}, 0 \le r \le 3;$

$f_{X_1,X_2}(r_1, r_2) = \binom{3}{r_1,\ r_2}\left(\frac{1}{3}\right)^3, r_1 \ge 0, r_2 \ge 0, r_1 + r_2 \le 3.$

7. Let A be the event that the first toss is heads. We find

	0	1	2	3	4
$f_{X\mid A}$	0	$\frac{1}{8}$	$\frac{3}{8}$	$\frac{3}{8}$	$\frac{1}{8}$
$f_{X\mid A'}$	$\frac{1}{8}$	$\frac{3}{8}$	$\frac{3}{8}$	$\frac{1}{8}$	0

10. **(a)** $\frac{2}{9}$, **(b)** $\frac{2}{9}$, **(c)** $\frac{7}{9}$, **(d)** $\frac{4}{9}$, **(e)** $\frac{2}{3}$.

	−2	0	1	2	3	4	6
f_{X+Y}	$\frac{1}{27}$	$\frac{2}{9}$	$\frac{1}{9}$	$\frac{2}{9}$	$\frac{1}{9}$	$\frac{5}{27}$	$\frac{1}{9}$

	−5	−2	−1	1	2	5
f_{XY}	$\frac{5}{27}$	$\frac{1}{9}$	$\frac{2}{9}$	$\frac{7}{27}$	$\frac{1}{9}$	$\frac{1}{9}$

The conditional probability that Y is odd given X is negative is $\frac{5}{8}$.

Section 6.2

4. Independent events in **(a)**, **(d)**, and **(e)**.
8. Independent events in **(a)**, **(c)**, and **(d)**.

Section 6.3

2. $\binom{30}{25}\left(\frac{1}{2}\right)^{30}, \sum_{j=0}^{25}\binom{30}{j}\left(\frac{1}{2}\right)^{30}.$

4. $\frac{1}{2}, \frac{1}{2}, \frac{93}{256}.$

6. $1 - (\frac{5}{6})^5 - (\binom{5}{1})(\frac{1}{6})(\frac{5}{6})^4.$

8. $\sum_{j=35}^{500}\binom{500}{j}\left(\frac{1}{20}\right)^j\left(\frac{19}{20}\right)^{500-j}.$

10. $\sum_{j=0}^{n}\left[\binom{n}{j}\frac{1}{2^n}\right]^2.$

12. $\frac{11}{32}$. **14.** $\frac{167}{256}$.

20. The first equation follows from noting that the event "At most k successes in n trials" may be broken up into the disjoint events "At most k successes in $n + 1$ trials" and "k successes in n trials and a success on the $(n + 1)$st trial."

Section 6.4

2. $\binom{8}{2,\,2,\,1,\,1,\,1}\left(\frac{1}{6}\right)^{8}$. **4.** $\binom{32}{16,\,7,\,8}\left(\frac{9}{16}\right)^{16}\left(\frac{3}{16}\right)^{15}\left(\frac{1}{16}\right)$.

6. $\sum_{j=0}^{3}\binom{10}{j}\left(\frac{2}{3}\right)^{j}\left(\frac{1}{3}\right)^{10-j}$. **8.** $\sum_{j=0}^{275}\binom{500}{j}\left(\frac{2}{5}\right)^{j}\left(\frac{3}{5}\right)^{500-j}$.

10. $\binom{20}{6,\,10}\left(\frac{4}{15}\right)^{6}\left(\frac{9}{50}\right)^{10}\left(\frac{83}{150}\right)^{4}$. **12.** $\sum_{j=11}^{20}\binom{20}{j}\left(\frac{2}{5}\right)^{j}\left(\frac{3}{5}\right)^{20-j}$.

Section 6.5

2. **(a)** $\frac{\binom{4}{2}\binom{48}{3}}{\binom{52}{5}}$ **(b)** $\sum_{j=2}^{4}\frac{\binom{4}{j}\binom{48}{5-j}}{\binom{52}{5}}$, **(c)** $\left(\frac{48}{52}\right)\left(\frac{47}{51}\right)$, **(d)** $1-\frac{\binom{46}{3}}{\binom{50}{3}}$,

(e) $\frac{\binom{13}{3}\binom{26}{2}+\binom{13}{2}\binom{26}{3}+\binom{13}{2}\binom{13}{1}\binom{26}{2}}{\binom{52}{5}}$, **(f)** $\frac{\binom{3}{1}\binom{48}{3}}{\binom{52}{5}}$, **(h)** $\frac{10\binom{4}{1}^{5}}{\binom{52}{5}}$,

(i) $\frac{156\binom{4}{2}\binom{4}{3}}{\binom{52}{5}}$, **(j)** $\frac{4}{\binom{52}{5}}$.

4. **(a)** $\sum_{j=7}^{12}\frac{\binom{35}{j}\binom{35}{12-j}}{\binom{70}{12}}$, **(b)** $\frac{\binom{45}{7}\binom{25}{5}}{\binom{70}{12}}$, **(c)** $\sum_{j=7}^{12}\frac{\binom{45}{j}\binom{25}{12-j}}{\binom{70}{12}}$,

(d) $\sum_{j=0}^{3}\frac{\binom{20}{j}\binom{25}{3-j}\binom{15}{5-j}\binom{10}{4+j}}{\binom{70}{12}}$,

(e) $\sum_{j=0}^{3}\frac{\binom{20}{j}\binom{25}{3-j}\binom{15}{5-j}\binom{10}{4+j}}{\binom{45}{3}\binom{25}{9}}$.

8. If $p=\sum_{j=4}^{6}\frac{\binom{8}{j}\binom{12}{6-j}}{\binom{20}{6}}$, then the probability that more than ten cartons are rejected is $\sum_{j=11}^{20}\binom{20}{j}p^{j}(1-p)^{20-j}$.

10. $\frac{\binom{10}{2}\binom{5}{2}\binom{3}{2}}{\binom{18}{6}}$.

Section 7.2

2. **(a)** $\frac{7}{2}$, **(b)** 7, **(c)** $\frac{161}{36}$, **(d)** $\frac{91}{36}$, **(e)** 0, **(f)** $\frac{70}{36}$, **(g)** $\frac{196}{36}$, **(h)** $\frac{1}{3}$, **(i)** $\frac{2}{3}$.
4. 2, 0.
6. The expectation on the draw is $\frac{105}{13}$ dollars.
8. Three dollars.
12. $\frac{185}{6}$ minutes, $\frac{1055}{36}$ minutes, at 25 minutes past the hour.
16. **(a)** 4, **(b)** $\frac{15}{2}$, **(c)** $\frac{27}{6}$, **(d)** 3, **(e)** $\frac{1}{2}$, **(f)** $\frac{3}{2}$, **(g)** $\frac{11}{2}$, **(h)** $\frac{1}{6}$, **(i)** $\frac{1}{3}$.
(a) $\frac{7}{2}$, **(b)** 7, **(c)** $\frac{7}{2}$, **(d)** $\frac{7}{2}$, **(e)** 0, **(f)** 0, **(g)** $\frac{7}{2}$, **(h)** $\frac{1}{3}$, **(i)** $\frac{2}{3}$.

Section 7.3

1. $E(X) = \frac{7}{2}$, $E(2X + 3) = 10$, $E(X^2) = \frac{91}{6}$, $E(3X^2 + 2X + 1) = \frac{107}{2}$, $E(1/X) = \frac{147}{360}$.
3. **(c)** $E(X) = \frac{5}{2}$, $E(Y) = \frac{31}{16}$, $E(XY) = \frac{47}{8}$, $E(X^2 + Y) = \frac{151}{16}$.

Section 7.4

2. **(a)** $\frac{12}{17}$, **(b)** $\frac{25}{68}$, **(c)** $\frac{169}{68}$, **(d)** $\frac{36}{17}$.
4. The variances are 1 and the covariance is -1.
6. Suppose the joint probability distribution of X and Y is

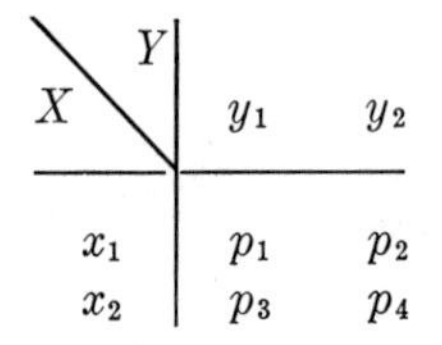

X \ Y	y_1	y_2
x_1	p_1	p_2
x_2	p_3	p_4

Write out $E(X)$, $E(Y)$, and $E(XY)$, and investigate the consequences of supposing that $E(XY) - E(X)E(Y) = 0$.

Section 7.5

2. p, pq/n.
4. **(a)** p, **(c)** p^2, **(d)** $np + n(n-1)p^2 + n(n-1)(n-2)p^3$.
5. $E(X_i) = np_i$, $V(X_i) = np_i(1 - p_i)$.
8. Cov $(X_1, X_2) = -np_1p_2$.
10. **(a)** $E(X_i) = (N+1)/2$, $V(X_i) = (N+1)(N-1)/12$,
(d) $f_{M_n}(k) = (k/N)^n - ((k-1)/N)^n$, $1 \le k \le N$,
$f_{m_n}(k) = ((N - k + 1)/N)^n - ((N-k)/N)^n$, $1 \le k \le N$.

Section 8.3

2. With replacement, the expectation is 1.05 and the variance is 219/400000. Without replacement, the expectation is 1.05 and the variance is 21681/39999600.

Section 8.4

2. $\sim .9774$.
4. $200 - 14$ and $200 + 14$.
6. Approximately, at least 952.
8. $P^n(S_n \geq \alpha) \cong 1 - \Phi\left(\frac{\alpha - np}{\sqrt{npq}}\right)$. **10.** $n \geq (49)^2$, $n \geq (980)^2$.
12. (i) .0336, (ii) .1112, (iii) .2709.

Section 8.5

2. .9483, .9999.
4. The approximate probability is less than .00005.
6. .0721. **8.** .6778.

Section 9.2

2. $(\frac{9}{2})e^{-3}$, $13e^{-3}$, $1 - (\frac{131}{8})e^{-3}$.
4. $s = 2$.

Section 9.3

2. A geometric distribution with $p = \frac{11}{36}$.
4. The expectation is 60, and the variance is 300.
6. The expectation is $1/(\sqrt{2} - 1)$.
8. $\sum_{k=0}^{n} \frac{e^{-\lambda}\lambda^{2k}}{(2k)!}$; this approaches $e^{-\lambda}(e^{\lambda} + e^{-\lambda})/2$ as $n \to \infty$.
11. $EX(X - 1)(X - 2) \cdots (X - n + 1) = \lambda^n$.

Index

C D E F G H I J
0 1 2 3 4 5 6